Vectors and Tensors in Crystallography

Vectors and Tensors in Crystallography

Donald E. Sands
Professor and Chairperson,
Department of Chemistry
University of Kentucky

DOVER PUBLICATIONS, INC.
NEW YORK

Copyright

Copyright © 1982 by Donald E. Sands.
All rights reserved under Pan American and International Copyright Conventions.

Published in Canada by General Publishing Company, Ltd., 30 Lesmill Road, Don Mills, Toronto, Ontario.
Published in the United Kingdom by Constable and Company, Ltd., 3 The Lanchesters, 162–164 Fulham Palace Road, London W6 9ER.

Bibliographical Note

This Dover edition, first published in 1995, is an unabridged, slightly corrected republication of the work first published by the Addison-Wesley Publishing Company (Advanced Book Program/World Science Division), Reading, Mass., 1982.

Library of Congress Cataloging-in-Publication Data

Sands, Donald, 1929–
Vectors and tensors in crystallography / Donald E. Sands. — Dover ed.
p. cm.
Originally published: Reading, Mass. : Addison-Wesley, 1982.
Includes bibliographical references and index.
ISBN 0-486-68505-5
1. Crystallography, Mathematical. 2. Vector analysis. 3. Tensor algebra. I. Title.
QD911.S24 1995
548′.7—dc20 94-40104
CIP

Manufactured in the United States of America
Dover Publications, Inc., 31 East 2nd Street, Mineola, N.Y. 11501

Contents

Preface

This treatment of vectors and tensors is intended especially for crystallographers, students of crystallography, and solid-state scientists, all of whom need proficiency in carrying out calculations in rectilinear coordinate systems. Few of the many excellent treatises of crystallography offer even a smattering of instruction on vector and tensor manipulations in general coordinate systems, and attempts to extend familiarity with cartesian vectors to rectilinear spaces are fraught with peril. There is an abundance also of fine books on vector and tensor analysis; to my knowledge, however, none of these includes the specific examples and applications that crystallographers need.

Readers with no special interest in crystallography may also find this work a useful introduction to vector and tensor analysis. The emphasis on rectilinear systems is greater than that found in conventional textbooks on the subject, but the crystallographic examples demonstrate the tremendous utility of tensors in describing the real physical world.

The book has been designed to be suitable for self-study or for use in a course. There are more than 300 exercises, and hints, suggestions, and answers are given in an appendix. It should be an appropriate textbook for a course in vector and tensor analysis, particularly if the principal interest is in rectilinear systems. It could serve also as the textbook for a course in crystallography, perhaps as a companion to a book that teaches the crystallographic principles, such as my own *Introduction to Crystallography*. Besides an understanding of the elements of crystallography, the main prerequisite for studying this book is some experience with matrix algebra; the reader without such background may wish to refer to an introductory exposition of matrices.

Whatever uniqueness of topic, content, and scope may be claimed for this work, a few other sources must be recognized that impinge upon its subject matter. Most crystallographers, for example, have at their fingertips Volume II of the *International Tables for X-Ray Crystallography,* which includes a twelve-page summary of tensor analysis by A. L. Patterson. W. H. Zachariasen's classic *Theory of X-Ray Diffraction in Crystals* contains a rigorous theoretical treatment of x-ray crystallography in which the tensor description is couched in dyadic notation. Another important book with some topical similarity to this one is J. F. Nye's *Physical Properties of Crystals,* which gives a lucid description of the electrical, magnetic, elastic, and other properties of crystals, all of which are handled with facility and elegance by means of tensors; Nye's treatment, though, is purely cartesian.

The core of the book may be regarded as Chapters 1, 3, and 4, which present the fundamentals of vector and tensor analysis in rectilinear systems. Chapter 1 develops the vector algebra of such systems. Some sensibilities may be offended by the abrupt appearance of such terms as covariant and contravariant in the very first section, or by references to metric tensors and permutation tensors long before tensors are defined; these deliberate deviations from a linear ordering of thought are intended to encourage correct notation from the beginning and to plant seeds of familiarity that may be exploited later.

Chapter 2 shows how lines and planes may be characterized. The treatment is brief, but it is an essential foundation for the chapters that follow.

The transformation theory in Chapter 3 shows what tensors are really all about. The principles are covered in nine sections, which are followed by eight sections of special applications. The general reader may not be interested in all these applications, and selectivity in perusal is encouraged. The section on derivatives makes the point that derivatives generate tensors in rectilinear systems, but not necessarily in curvilinear systems; this fine distinction may clarify the explanations, but it is not crucial to the evolvement of the overall subject. The reader whose goal is to learn tensor analysis will be forgiven for omitting the abstruse section on rigid-body motion. The section on dyadics was added at the behest of a reviewer of the first draft of the manuscript.

Chapter 4 presents a tensor treatment of symmetry. The discussion covers both point groups (in direct and reciprocal space) and space groups. Since the aim is to illuminate principles, a methodical derivation of all crystallographic groups is not attempted.

My original outline did not include the physical properties of crystals. The fine books by Nye and others seemed to fill that need admirably. Chapter 5 was added later, though, in response to a reviewer's urging. In keeping with my purpose of teaching tensor analysis, I have avoided cataloging the various anisotropic properties of crystals or reviewing the state of research in these fields. Instead, the chapter provides a few examples of tensor properties, shows how to handle them mathematically, and touches on generalizations to noncartesian systems.

Chapter 6 deals with some specific experimental techniques for observing the diffraction of x rays by crystals. A unifying theme of this chapter is the description in terms of two cartesian coordinate systems, one fixed in space and one attached to the crystal. Few readers will want to suffer through the details of all the methods, and much of this material will be irrelevant to the crystallographer interested only in the numbers produced by a fully automated data collection system. Again, the intention is to teach principles, not to prepare an operating manual for instruments, and the important features of this chapter are the concepts that can be applied to other experimental arrangements.

Chapter 7 is a brief survey of curvilinear spaces, which is meant to clarify by generalization some of the material developed in the preceding chapters and, above all, to lay the foundation for further study.

It was inevitable that notational problems would be encountered in assembling this book. Ideally, different quantities should be represented by different symbols, but it is desirable also to adhere to conventional usage. As an example of the difficulties that appear, the symbol σ is used variously to denote a reciprocal lattice vector, the electrical conductance tensor, and the stress tensor. The context should reveal the proper meaning in all cases, but caution should be observed in interpreting all symbols. Confusion may occur also between the superscript used as a contravariant index and the exponent indicating raising a quantity to a power; again, the meaning should be clear from the context. A different kind of dilemma arose in the treatment of kappa geometry, where the sign conventions applied to all the other experimental techniques in Chapter 6 were opposite to those used by the manufacturer of the commercial diffractometer that uses kappa geometry; either choice was bound to confuse, but I elected to follow the uniform sign conventions of the preceding sections rather than to accommodate to the design of a particular instrument.

The development of this book owes much to the suggestions of two anonymous reviewers who inspected an early version of the manuscript. I am grateful for the valuable comments of Dr. Carolyn P. Brock, who helped to prevent several blunders in Chapter 6. I thank also Mrs. Eula Moore for assistance with the preparation of the final manuscript, Mrs. Barbara Mabry for reading the manuscript with the critical eye of an English teacher, Mr. Douglas K. Cole for converting my sketches to finished drawings, and the staff of Addison-Wesley Publishing Company, Advanced Book Program/World Science Division, for their patience and encouragement.

DONALD E. SANDS

Glossary of Symbols

This glossary lists many of the symbols used in the book. Obvious symbols or those that are used infrequently are not listed here. Sometimes different quantities are represented by the same symbol, so the meaning of each symbol should be determined carefully from the local context.

$\mathbf{a}_i$	Covariant basis vector
$\mathbf{a}^i$	Contravariant basis vector
$\mathbf{A}$	Matrix with component A^{ij}, $A^i{}_j$, $A_i{}^j$, or A_{ij} in row i and column j
$\bar{\mathbf{A}}$	Transpose of matrix **A**: component A^{ij} is in row j, column i
$\lvert\mathbf{A}\rvert$	Determinant of square matrix **A**, usually just given as A
B	Isotropic temperature factor
B	Goniometer head arc nearer the base
$\mathbf{c}$	Elastic stiffness tensor with components c^{ijkl}
c^{mn}	Two-index elastic stiffness components
$\mathbf{C}_n$	Schoenflies rotation operator
d	Interplanar spacing (equal to $1/h$)
$\mathbf{e}_i$	Cartesian basis vector
e_{ijk}	Permutation symbol

F	Transformation matrix for contravariant quantities: $x'^i = F^i{}_j x^j$
G	Transformation matrix for covariant quantities: $a'_i = G_i{}^j a_j$
g	Metric tensor with components g_{ij}
g	Determinant of **g** (equal to the square of V)
g_{ij}	Component of **g**, equal to $\mathbf{a}_i \cdot \mathbf{a}_j$
g*	Reciprocal metric tensor with components g^{ij}
g^*	Determinant of **g***
g^{ij}	Component of **g***, equal to $\mathbf{a}^i \cdot \mathbf{a}^j$
h	Reciprocal lattice vector with components h_i
h	Length of vector **h**
h_i	Covariant component of vector **h**; the conventional Miller indices, h, k, l, are related to the h_i by $h_1 = h$, $h_2 = k$, $h_3 = l$
I	Unit matrix with components $\delta^i{}_j$
j	Current density
J	Jacobian determinant of a transformation, equal to $\lvert\mathbf{G}\rvert$
L	Lorentz factor
L	Libration tensor with components L^{ij}
p	Polarization factor
P	Parity operator, equal to the sign of $\mathbf{a}_1 \cdot \mathbf{a}_2 \wedge \mathbf{a}_3$
P(u/v)	Projection of vector **u** onto vector **v**
P(u‖h)	Projection of vector **u** parallel to **h** onto the plane normal to **h**
R	Rotation operator with components $R^i{}_j$
R	Projection of **s** onto the plane normal to the precession axis
$\mathbf{R}_0$	Projection of $\mathbf{s}_0$ onto the plane normal to the precession axis
$\mathbf{s}_0$	Vector of length $1/\lambda$ in the direction of the incident x-ray beam
s	Vector of length $1/\lambda$ in the direction of the diffracted x-ray beam
s	Elastic compliance tensor with components s_{ijkl}
s_{mn}	Two-index elastic compliance components
S	Screw tensor with components S^{ij}
S	Symmetry operator with components $S^i{}_j$
$\mathbf{S}_n$	Schoenflies improper rotation operator
T	Goniometer head arc at the top, nearer the crystal
T	Translation tensor with components T^{ij}
U	Anisotropic temperature factor tensor with components $U^{ij} = \langle u^i u^j \rangle = \beta^{ij}/2\pi^2$

$\mathbf{U}$	Matrix of $U_{ij} = \partial u_i/\partial x^j$, where u_i = displacement
$\mathbf{U}$	Orientation matrix
$\mathbf{v}$	A vector
v	The length of vector $\mathbf{v}$
v^i	Contravariant component of vector $\mathbf{v} = [v^1, v^2, v^3]$
v_i	Covariant component of vector $\mathbf{v} = (v_1, v_2, v_3)$
V	Unit cell volume
V^*	Volume of reciprocal cell $= 1/V$
$\mathbf{X}$	Radial component of vector $\mathbf{h}$ (normal to rotation axis)
$\mathbf{Z}$	Component of $\mathbf{h}$ along rotation axis
α	Angle between $\mathbf{a}_2$ and $\mathbf{a}_3$
α	Inclination of $\varkappa$ angle to $\mathbf{e}_3$
α	Thermal expansion tensor with components α_{ij} or $\alpha^i{}_j$
α^*	Angle between $\mathbf{a}^2$ and $\mathbf{a}^3$
β	Angle between $\mathbf{a}_1$ and $\mathbf{a}_3$
β^{ij}	Anisotropic temperature factor component in exp $(-\beta^{ij}h_ih_j)$
β^*	Angle between $\mathbf{a}^1$ and $\mathbf{a}^3$
γ	Angle between $\mathbf{a}_1$ and $\mathbf{a}_2$
γ^*	Angle between $\mathbf{a}^1$ and $\mathbf{a}^2$
$\delta^i{}_j$	Kronecker delta = 1 if $i = j$, 0 if $i \neq j$
ϵ	Strain tensor with components ϵ_{ij}
ϵ_{ijk}	Component of permutation tensor $= PVe_{ijk}$
ϵ^{ijk}	Component of permutation tensor $= PV^*e_{ijk}$
ζ	Cylindrical coordinate equal to λZ
η	Supplement of angle between $\mathbf{X}$ and $\mathbf{R}_0$ in precession geometry
θ	Bragg angle
θ	Spherical coordinate
$\varkappa$	Kappa diffractometer angle
λ	Wavelength of radiation
λ	Eigenvalue of tensor
μ	Angle between precession axis and $\mathbf{s}_0$ in precession geometry
υ	Complement of angle between $\mathbf{s}$ and rotation axis in rotation or Weissenberg geometry; angle between $\mathbf{s}$ and precession axis in precession geometry
ξ	Cylindrical coordinate equal to λX
σ	Reciprocal lattice vector multiplied by wavelength, equal to $\lambda\mathbf{h}$
σ	Electrical conductivity tensor

σ Stress tensor

τ Angular cylindrical coordinate of **h**

Υ Angle between $\mathbf{s}_0$ and projection of **s** onto $\mathbf{e}_1$–$\mathbf{e}_2$ plane in rotation or Weissenberg geometry; angle between **R** and $\mathbf{R}_0$ in precession geometry

ϕ Diffractometer angle

ϕ Spherical coordinate

Φ A dyadic

χ A parameter in precession geometry given by $\cos \chi = \sin \mu \sin\omega$

χ Diffractometer angle in Eulerian geometry

ψ Angle of rotation about **h**

ω Diffractometer angle

ω Rotation about $\mathbf{s}_0$ in precession geometry

$\mathbf{u} \cdot \mathbf{v}$ Scalar product of vectors **u** and **v**

$\mathbf{u} \Lambda \mathbf{v}$ Vector product of vectors **u** and **v**

∇ Del operator

$\nabla\psi$ Gradient of ψ

$\nabla \cdot \mathbf{v}$ Divergence of vector **v**

$\nabla \Lambda \mathbf{v}$ Curl of vector **v**

∇^2 Laplacian operator

$v^i{}_{,j}$ Covariant derivative of v^i

$v_{i,j}$ Covariant derivative of v_i

$[ij, k]$ Christoffel symbol of first kind

$\left\{ {k \atop ij} \right\}$ Christoffel symbol of second kind

$\{\mathbf{R}|\mathbf{t}\}$ Space group operator: $\{\mathbf{R}|\mathbf{t}\}\,\mathbf{v} = \mathbf{R}\mathbf{v} + \mathbf{t}$

Vectors and Tensors in Crystallography

CHAPTER 1

Vectors

1-1 Introduction

Vectors have both magnitude and direction. Quantities that have magnitude only are called *scalars*; examples are the temperature of an object, or its mass, or its volume, or its length, each of which may be represented by a single number. A temperature gradient, on the other hand, is a *vector* quantity; its complete description requires statements of the rate of change of the temperature (in degrees per unit length) along each of three independent directions. Likewise, electric field strength is a vector quantity.

The thermal conductivity of an object is an example of an even higher set of quantities called *tensors*; the tensor in this case shows how each of the components of the vector representing the flow of heat varies with each of the components of the thermal gradient. It is convenient to regard scalars and vectors as special cases, of zero and first rank respectively, of the general subject of tensors.

This chapter treats vectors in general rectilinear coordinate systems. A mathematical expression of such a vector is

$$\mathbf{v} = x\mathbf{a} + y\mathbf{b} + z\mathbf{c} \tag{1-1}$$

The quantities x, y, z are numbers that represent the *components* of the vector $\mathbf{v}$; that is, x, y, z are the coordinates of the point reached from the origin by $\mathbf{v}$. The quantities $\mathbf{a}$, $\mathbf{b}$, $\mathbf{c}$ are themselves vectors that define the space or coordinate system; $\mathbf{a}$, $\mathbf{b}$, $\mathbf{c}$ are known as the *basis vectors* of the space. The term *rectilinear* means that the basis vectors have fixed directions and lengths, in contrast to *curvilinear* systems, in which the directions and lengths of the basis vectors may change from point to point.

A special case of rectilinear space is *cartesian* space, in which the basis vectors are orthonormal (mutually perpendicular and each of unit length). Familiarity with cartesian vector algebra may make some of what is to come easier, but the development of the subject will be reasonably self-sufficient.

The two-dimensional example depicted in Fig. 1-1 illustrates how Eq. (1-1) represents a vector. The vector with components $x = 0.6$, $y = 1.8$ is generated by going from the origin distance $0.6a$ (where a represents the length of $\mathbf{a}$) along the $\mathbf{a}$ axis, then moving a distance

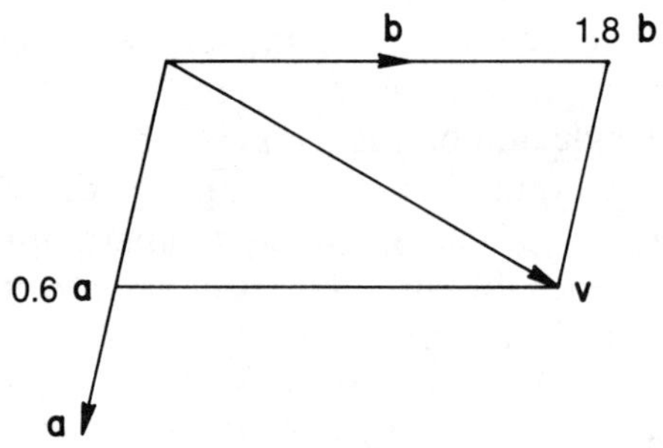

FIGURE 1-1. The vector $\mathbf{v} = 0.6\mathbf{a} + 1.8\mathbf{b}$

1.8b parallel to **b**. If there is a third component, we move finally zc parallel to **c**.

When we refer to rectilinear versus curvilinear or cartesian space, we mean, of course, merely the manner in which we choose to describe a particular problem; the nature of the real physical world is independent of our preferences for doing calculations.

Considerable economy of notation may be achieved by replacing the components x, y, z of vector **v** by the symbols v^1, v^2, v^3, and the basis vectors **a**, **b**, **c** by $\mathbf{a}_1$, $\mathbf{a}_2$, $\mathbf{a}_3$. The distinction between superscripts and subscripts will be important when we take up transformations in Chapter 3, so we shall start adjusting to proper usage right away. Quantities with subscripts are said to be *covariant* (remember that co- is low), whereas superscripts indicate *contravariant* quantities. With this notation, Eq. (1–1) becomes

$$\mathbf{v} = \Sigma v^i \mathbf{a}_i \tag{1–2}$$

where the summation is for $i = 1, 2, 3$. Vector and tensor equations in which an index appears once as a subscript and once as a superscript more often than not involve summation over that index. It is therefore convenient to omit the summation sign in Eq. (1–2) and write simply

$$\mathbf{v} = v^i \mathbf{a}_i \tag{1–3}$$

where summation over i is implied by the repeated index. This is known as the *Einstein summation convention,* and we shall observe this convention throughout the remainder of this book. A lower-case letter that is repeated in a covariant and a contravariant position will imply summation unless we state explicitly that no summation is intended in a particular case. Thus, an expression such as $A^{ij}{}_k\, B^k{}_{lm}$ is shorthand notation for $A^{ij}{}_1\, B^1{}_{lm} + A^{ij}{}_2\, B^2{}_{lm} + A^{ij}{}_3\, B^3{}_{lm}$. The repeated index may of course be replaced by any other symbol: $A^{ij}{}_k\, B^k{}_{lm}$ is no different from $A^{ij}{}_n\, B^n{}_{lm}$. The repeated index, therefore, is really a dummy index for which we can substitute any other letter that is not used elsewhere in the expression or equation. Sometimes, repeated indices without intended summation are designated by capital letters; for example, $v^J \mathbf{a}_J$ represents a single term.

We shall denote cartesian basis vectors by $\mathbf{e}_i$ and the corresponding components by l^i. In terms of cartesian components the vector $\mathbf{v}$ may be expressed as

$$\mathbf{v} = l^i \mathbf{e}_i \tag{1-4}$$

1-2 Lattice Vectors

Much of our development of vector and tensor analysis, and many of our applications, will come from the field of crystallography. An excellent definition of a crystal is that given by Barrett (1952): "A crystal consists of atoms or molecules arranged in a pattern that repeats periodically in three dimensions." An elementary exposition of the basic principles of crystallography has been presented in an earlier book (Sands, 1969), to which the reader is referred for a more descriptive and less mathematical treatment than that given here. This book, in fact, complements the previous one, filling in gaps in rigor and completeness.

Figure 1-2 shows a two-dimensional periodic arrangement. To aid in describing such a structure, we select a set of reference points by choosing any point in the pattern and marking all points that are translationally indistinguishable. This set of indistinguishable points is the set of *lattice points*. It is important to understand that the selection of the first point is completely arbitrary; it is the indistinguishability of the points from each other that characterizes them as lattice points. Some choices of lattice points may be more convenient than others—at, for example, certain atoms, or at points that are uniquely defined by elements of rotational or reflectional symmetry—but the lattice concept itself requires only a periodic array. Figure 1-2 demonstrates the need for translational indistinguishability; points X and Y are at symmetry-related positions, but they are not translationally the same: If you slide from X to Y, you can presumably tell that the nearest motif is situated differently.

Having selected the lattice points, we proceed to connect them to outline parallelepipeds—the unit cells of the structure. The unit cell serves as a template for the structure; translation of a single unit cell from one lattice point to another may be thought of as generating the entire crystal. To describe a crystal structure we need specify only the locations of the atoms within a single unit cell.

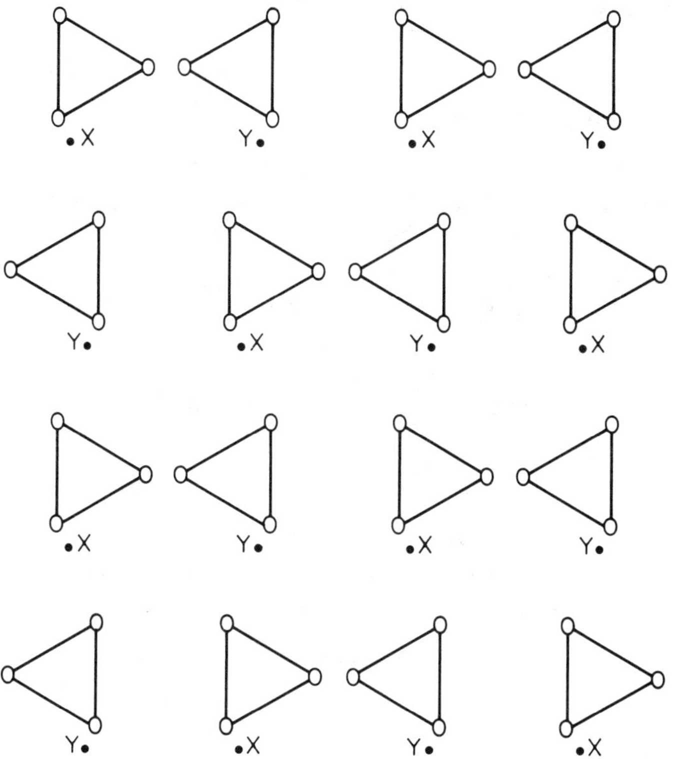

FIGURE 1-2. A periodic structure. Points X and Y are equivalent by symmetry, but not translationally indistinguishable.

The vectors from one lattice point chosen to be the origin to other lattice points are the *lattice vectors*. Any three noncoplanar lattice vectors define a unit cell. Since an infinite crystal possesses an infinite number of lattice vectors, there is nothing unique about the choice of unit cell. The presence of elements of point symmetry, however, imposes conditions upon the lattice geometry, and symmetry provides a basis for the classification of crystals into systems with characteristic restrictions upon the unit cell dimensions.

The spatial distribution of lattice points in a given crystal depends upon the manner in which the constituent atoms or molecules are packed together. Nature establishes the lattice geometry, and it often is to our advantage to describe it in nature's terms. The vector algebra and theory of transformations of nonorthogonal geometry are thus essential tools of crystallography.

1-3 Addition and Subtraction of Vectors

The sum of two vectors **u** and **v** is the vector obtained by going along the direction of **u** a distance u and then going parallel to **v** a distance v. The resultant vector **w** is given by

$$\mathbf{w} = \mathbf{u} + \mathbf{v} = \mathbf{v} + \mathbf{u} \tag{1-5}$$

Since $(-\mathbf{v})$ is a vector of length v pointing opposite to **v**, subtraction of **v** from **u** is equivalent to adding $(-\mathbf{v})$ to **u**; thus,

$$\mathbf{t} = \mathbf{u} - \mathbf{v} = \mathbf{u} + (-\mathbf{v}) \tag{1-6}$$

EXERCISE 1-1 Draw to scale a two-dimensional coordinate system for which $a_1 = 5.00$, $a_2 = 8.00$ Å, and ϕ, the angle between $\mathbf{a}_1$ and $\mathbf{a}_2$, is 50.0°. Draw the vectors **u** and **v**, where $\mathbf{u} = [0.50, 1.00]$, $\mathbf{v} = [0.75, -0.25]$.

EXERCISE 1-2 Calculate the components of $\mathbf{u} + \mathbf{v}$ and $\mathbf{u} - \mathbf{v}$ for the vectors of Exercise 1-1.

1-4 Scalar Product

Carefully defined operations analogous to arithmetic multiplication may be applied to vectors. One such operation produces a scalar quantity and is called the *scalar product*; it is also called the *dot product* or the *inner product*, and it is symbolized by $\mathbf{u} \cdot \mathbf{v}$ or sometimes $(\mathbf{uv})$. The scalar product of vectors **u** and **v** is defined as

$$\mathbf{u} \cdot \mathbf{v} = uv \cos \phi \tag{1-7}$$

where ϕ is the angle between **u** and **v**. In terms of components,

$$\mathbf{u} \cdot \mathbf{v} = u^i v^j \mathbf{a}_i \cdot \mathbf{a}_j \tag{1-8}$$

EXERCISE 1-3 If $u = 5.00$, $v = 7.00$ Å, and $\phi = 50.00°$, what is the value of $\mathbf{u} \cdot \mathbf{v}$?

EXERCISE 1–4 Show that $\mathbf{e}_i \cdot \mathbf{e}_j = \delta_{ij}$, where $\mathbf{e}_i$ is a cartesian basis vector and δ_{ij} is the Kronecker delta, which is equal to 1 if $i = j$ and equal to 0 if $i \neq j$.

1–5 Length of a Vector

When the two vectors in the scalar product are the same, $\phi = 0$, $\cos \phi = 1$, and Eq. (1–7) becomes

$$\mathbf{v} \cdot \mathbf{v} = v^2 \tag{1–9}$$

This equation provides a means of calculating the length of a vector. Using Eq. (1–3),

$$\mathbf{v} \cdot \mathbf{v} = (v^i \mathbf{a}_i) \cdot (v^j \mathbf{a}_j) \tag{1–10}$$

which is equivalent to

$$\mathbf{v} \cdot \mathbf{v} = v^i v^j \mathbf{a}_i \cdot \mathbf{a}_j \tag{1–11}$$

Equation (1–11) is shorthand for nine terms: With the conventions that α is the angle between $\mathbf{a}_2$ and $\mathbf{a}_3$, β the angle between $\mathbf{a}_1$ and $\mathbf{a}_3$, and γ the angle between $\mathbf{a}_1$ and $\mathbf{a}_2$, Eq. (1–11) expands to

$$\begin{aligned}\mathbf{v} \cdot \mathbf{v} = {} & (v^1 a_1)^2 + (v^2 a_2)^2 + (v^3 a_3)^2 + \\ & 2v^1 v^2 a_1 a_2 \cos \gamma + 2v^1 v^3 a_1 a_3 \cos \beta + \\ & 2v^2 v^3 a_2 a_3 \cos \alpha\end{aligned} \tag{1–12}$$

EXERCISE 1–5 Calculate the lengths of each of the vectors $\mathbf{u}$ and $\mathbf{v}$ of Exercise 1–1.

EXERCISE 1–6 Calculate the lengths of each of the vectors $\mathbf{u} + \mathbf{v}$ and $\mathbf{u} - \mathbf{v}$ of Exercise 1–2.

1–6 Angle between Two Vectors

Equations (1–7) and (1–8) also provide a means of finding the angle between two vectors. Since the lengths of vectors can be determined from Eq. (1–12), ϕ is obtained from

$$\cos \phi = u^i v^j \mathbf{a}_i \cdot \mathbf{a}_j / (uv) \tag{1–13}$$

EXERCISE 1-7 If $u = 4.00$, $v = 7.00$, and $\mathbf{u} \cdot \mathbf{v} = 15.00$, what is the angle between **u** and **v**?

EXERCISE 1-8 Calculate the angle between the vectors $\mathbf{u} + \mathbf{v}$ and $\mathbf{u} - \mathbf{v}$ of Exercise 1-2.

EXERCISE 1-9 Show that the calculation of the length of vector $\mathbf{s} - \mathbf{t}$ leads to the trigonometric law of cosines.

EXERCISE 1-10 Calculate the angle between the body diagonals of a cube by applying Eq. (1-7) to the cartesian vectors [1, 1, 1] and [1, −1, −1].

EXERCISE 1-11 Calculate the angle between the body diagonal [1, 1, 1] and the face diagonal [1, −1, 0] of a cube.

EXERCISE 1-12 Calculate the angle between the face diagonals [1, 1, 0] and [0, −1, 1] of a cube.

EXERCISE 1-13 Find a vector of unit length that is perpendicular to both [1, 1, 0] and [1, 1, 1] in a cubic crystal.

EXERCISE 1-14 Suppose a coordinate system has $a_1 = 5.00$, $a_2 = 7.00$, $a_3 = 9.00$ Å, $\alpha = 120.00°$, $\beta = 85.00°$, $\gamma = 75.00°$.

(a) Calculate the length of the vector with components [0.200, −0.300, 0.600].

(b) Calculate the length of the vector with components [0.500, 0.700, −0.100].

(c) Calculate the angle between the vectors of (a) and (b).

(d) Find a vector **v** such that it forms a triangle with the vectors of (a) and (b). What is the length of **v**?

EXERCISE 1-15 Calculate the angles between the vector with components [0.200, −0.300, 0.600] and the axes of the coordinate system in Exercise 1-14.

EXERCISE 1-16 Prove vectorially that the diagonals of a rhombus are perpendicular to each other.

1-7 Metric Tensor

Scalar products of basis vectors, $\mathbf{a}_i \cdot \mathbf{a}_j$ occur in numerous formulas, and for computational purposes these scalar products are of greater utility than the vectors $\mathbf{a}_i$ themselves or the angles between them. We write

$$g_{ij} = \mathbf{a}_i \cdot \mathbf{a}_j \qquad (1\text{-}14)$$

where the g_{ij} quantities are referred to as the components of the metric tensor. The word "metric" is used to indicate that the metric tensor pertains to the measurement properties of the space; specifically, the quantities g_{ij} contain information about the lengths of the basic vectors and the angles between them. The word "tensor" implies that the g_{ij} transform in a certain way upon choosing a new set of basis vectors; these transformation properties are the subject of Chapter 3 and will not be needed here, but the term is introduced now in order to give the g_{ij} a proper name. Although this is our first encounter with a tensor (except for scalars and vectors, which, as we shall see, are special cases of tensors), we shall become very familiar with the metric tensor before we get around to a formal definition.

Since

$$g_{ij} = g_{ji} \tag{1-15}$$

the metric tensor is said to be symmetric, so it has six independent components in three dimensions: g_{11}, g_{12}, g_{13}, g_{22}, g_{23}, and g_{33}.

In terms of the metric tensor, the scalar product of two vectors **u** and **v** is

$$\mathbf{u} \cdot \mathbf{v} = u^i v^j g_{ij} \tag{1-16}$$

This formula for a scalar product is quite amenable to computer calculation. For example, the Fortran instructions required to calculate $\mathbf{u} \cdot \mathbf{v} =$ UV are

```
    DIMENSION U(3), V(3), G(3,3)
    UV = 0.0
    DO 5 I = 1, 3
    DO 5 J = 1, 3
  5 UV = UV + U(I) * V(J) * G(I,J)
```

The matrix representation of the metric tensor is useful also:

$$\mathbf{g} = \begin{pmatrix} g_{11} & g_{12} & g_{13} \\ g_{21} & g_{22} & g_{23} \\ g_{31} & g_{32} & g_{33} \end{pmatrix} \tag{1-17}$$

The reader who is unfamiliar with matrices will find it helpful to refer to one of the many available excellent introductions to matrix algebra (e.g., Ayres, 1962). Here, g_{ij} is the element in the ith row and jth column of $\mathbf{g}$. In matrix language,

$$\mathbf{u} \cdot \mathbf{v} = \bar{\mathbf{u}}\mathbf{g}\mathbf{v} = \bar{\mathbf{v}}\mathbf{g}\mathbf{u} \tag{1-18}$$

where

$$\mathbf{u} = \begin{pmatrix} u^1 \\ u^2 \\ u^3 \end{pmatrix}, \quad \mathbf{v} = \begin{pmatrix} v^1 \\ v^2 \\ v^3 \end{pmatrix} \tag{1-19}$$

and a bar over the symbol for a matrix indicates the transpose obtained by exchanging rows and columns:

$$\bar{\mathbf{u}} = [u^1, u^2, u^3], \qquad \bar{\mathbf{v}} = [v^1, v^2, v^3] \tag{1-20}$$

EXERCISE 1-17 Calculate the metric tensor components for the coordinate system of Exercise 1-14.

EXERCISE 1-18 A metric tensor has components $g_{11} = 57.2$, $g_{12} = 0$, $g_{13} = -4.20$, $g_{22} = 75.0$, $g_{23} = 0$, $g_{33} = 110.0$. Calculate the lengths of the basis vectors and the angles between them.

EXERCISE 1-19 Verify that by the rules of matrix multiplication (if $\mathbf{C} = \mathbf{AB}$, then $C_i{}^j = A_{ik}B^{kj}$) Eq. (1-18) is equivalent to Eq. (1-16).

EXERCISE 1-20 Find the angle between the vectors $\mathbf{u} = [0.320, -0.105, 0.200]$ and $\mathbf{v} = [0.165, 0.302, -0.080]$ in a coordinate system with $a_1 = 6.50$, $a_2 = 8.86$, $a_3 = 7.25$ Å, $\alpha = 71.2°$, $\beta = 108.5°$, $\gamma = 90.0°$.

EXERCISE 1-21 A crystal has $a_1 = 5.00$, $a_2 = 7.00$, $a_3 = 9.00$ Å, $\alpha = 75.0°$, $\beta = 85.0°$, $\gamma = 120.0°$. Use the metric tensor to calculate the angle between the vectors $[0.200, -0.300, 0.600]$ and $[0.500, 0.700, -0.100]$.

EXERCISE 1-22 If there are atoms at the points $[0.100, 0.600, 0.050]$ and $[0.900, 0.400, -0.050]$ of the crystal of Exercise 1-21, calculate the bond length (the shortest interatomic distance, even when periodicity is considered).

EXERCISE 1-23 ScF_3 is rhombohedral with $a_1 = 4.023$ Å, $\alpha = 89°34'$. A Sc atom is at 0, 0, 0, and the three F atoms are at 0.500, 0.027, 0.973; 0.973, 0.500, 0.027; and 0.027, 0.973. 0.500. Calculate the Sc–F bond length, the shortest F–F distances, and the F–Sc–F angles.

1-8 Vector Product

A second multiplicative operation, valid in three-dimensional space, is the *vector product* (or cross product) of two vectors. The vector product of **u** and **v**, written $\mathbf{u} \wedge \mathbf{v}$ (or $\mathbf{u} \times \mathbf{v}$, or $[\mathbf{uv}]$) is defined as a vector of length $uv \sin \phi$ in the direction perpendicular to **u** and **v** and in the sense such that **u**, **v**, and $\mathbf{u} \wedge \mathbf{v}$ form a right-handed system. Given **u** and **v**, we generate $\mathbf{u} \wedge \mathbf{v}$ by rotating **u** into **v** and advancing in the direction of a right-handed screw (see Fig. 1-3). The quantity $\mathbf{u} \wedge \mathbf{v}$ sometimes is referred to as an axial vector; axial vectors are associated with rotations, whereas regular (or polar) vectors involve only translations. An intrinsic difference between axial and polar vectors will be explored in Section 4-11.

From the definition, it follows that

$$\mathbf{v} \wedge \mathbf{u} = -\mathbf{u} \wedge \mathbf{v} \tag{1-21}$$

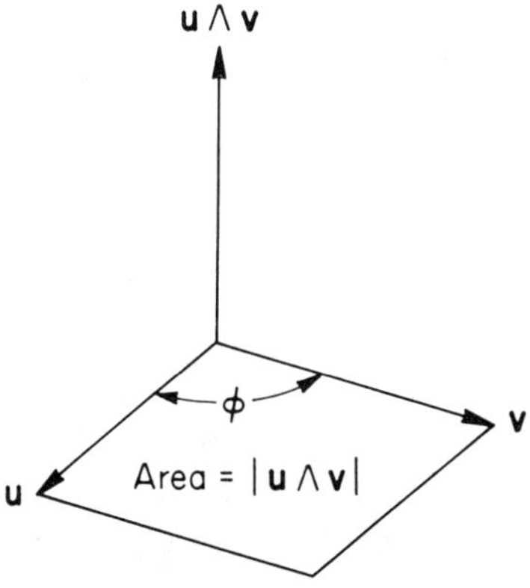

FIGURE 1-3. Vector product $\mathbf{u} \wedge \mathbf{v}$.

since a right-handed screw that rotates **v** into **u** points in the opposite direction of a right-handed screw that rotates **u** into **v**.

The magnitude of $\mathbf{u} \wedge \mathbf{v}$, $uv \sin \phi$, is numerically equal to the area of the parallelogram defined by the vectors **u** and **v**. This is illustrated in Fig. 1-3.

EXERCISE 1-24 Evaluate $\mathbf{e}_i \wedge \mathbf{e}_j$ for all i and j, where $\mathbf{e}_1$, $\mathbf{e}_2$, and $\mathbf{e}_3$ are cartesian basis vectors. Collect the results in a 3×3 array (matrix) with $\mathbf{e}_i \wedge \mathbf{e}_j$ in row i and column j.

EXERCISE 1-25 If **r** and **s** are vectors referred to cartesian space, show that $\mathbf{r} \wedge \mathbf{s}$ is given by the determinant

$$\begin{vmatrix} \mathbf{e}_1 & \mathbf{e}_2 & \mathbf{e}_3 \\ r^1 & r^2 & r^3 \\ s^1 & s^2 & s^3 \end{vmatrix}$$

EXERCISE 1-26 Evaluate $\mathbf{u} \cdot (\mathbf{u} \wedge \mathbf{v})$. Under what conditions is $\mathbf{u} \cdot \mathbf{v} \wedge \mathbf{w}$ equal to zero?

1-9 Triple Scalar Product

The quantity $\mathbf{u} \cdot \mathbf{v} \wedge \mathbf{w}$ is called the triple scalar product of the vectors **u**, **v**, and **w**. The vector product of **v** and **w** must be calculated first, then the scalar product of **u** with the vector $\mathbf{v} \wedge \mathbf{w}$ is calculated. That is,

$$\mathbf{u} \cdot \mathbf{v} \wedge \mathbf{w} = \mathbf{u} \cdot (\mathbf{v} \wedge \mathbf{w}) \tag{1-22}$$

but it is not necessary to include the parentheses, since $(\mathbf{u} \cdot \mathbf{v}) \wedge \mathbf{w}$ implies the meaningless operation of taking the vector product of the scalar product $\mathbf{u} \cdot \mathbf{v}$ with the vector **w**.

The volume of a parallelepiped is given by the area of the base times the altitude. The area of the parallelogram generated by vectors **v** and **w** is $|\mathbf{v} \wedge \mathbf{w}|$. The vector $\mathbf{v} \wedge \mathbf{w}$ is normal to this parallelogram. If **n** is a unit vector in the direction of $\mathbf{v} \wedge \mathbf{w}$, $\mathbf{u} \cdot \mathbf{n}$ is the length of the projection of **u** onto the vector $\mathbf{v} \wedge \mathbf{w}$, which is to say that $\mathbf{u} \cdot \mathbf{n}$ is the altitude of the parallelepiped with base defined by **v** and **w**. The volume of the parallelepiped is therefore $\mathbf{u} \cdot \mathbf{n}|\mathbf{v} \wedge \mathbf{w}|$, or

$$V = |\mathbf{u} \cdot \mathbf{v} \wedge \mathbf{w}| \tag{1-23}$$

These vectors are depicted in Fig. 1-4. The absolute value signs in Eq. (1-23) permit association of the triple scalar product with volume, even in cases of left-handed triplets for which $\mathbf{u} \cdot \mathbf{v} \wedge \mathbf{w}$ is negative.

There is nothing unique about our labeling of the vectors, so

$$\begin{aligned} \mathbf{u} \cdot \mathbf{v} \wedge \mathbf{w} &= \mathbf{v} \cdot \mathbf{w} \wedge \mathbf{u} = \mathbf{w} \cdot \mathbf{u} \wedge \mathbf{v} \\ &= -\mathbf{u} \cdot \mathbf{w} \wedge \mathbf{v} = -\mathbf{v} \cdot \mathbf{u} \wedge \mathbf{w} = -\mathbf{w} \cdot \mathbf{v} \wedge \mathbf{u} \end{aligned} \tag{1-24}$$

It should be apparent that V is zero if any two of the vectors are the same or parallel to each other (see Exercise 1-26).

A more useful formula for calculating the volume of a unit cell is

$$\begin{aligned} V = a_1 a_2 a_3 (1 &- \cos^2 \alpha - \cos^2 \beta - \cos^2 \gamma \\ &+ 2 \cos \alpha \cos \beta \cos \gamma)^{1/2} \end{aligned} \tag{1-25}$$

This may be derived by straightforward geometry (see Buerger, 1942); in Section 3-5 we shall obtain it rather simply from transformation theory. The formula is given here without proof so that it will be available for some of our calculations and for some forthcoming developments.

EXERCISE 1-27 A monoclinic unit cell has $a_1 = 5.00$, $a_2 = 7.00$, $a_3 = 8.50$ Å, $\alpha = 90.0°$, $\beta = 108.0°$, $\gamma = 90.0°$. Use Eq. (1-23) to calculate the volume of this unit cell.

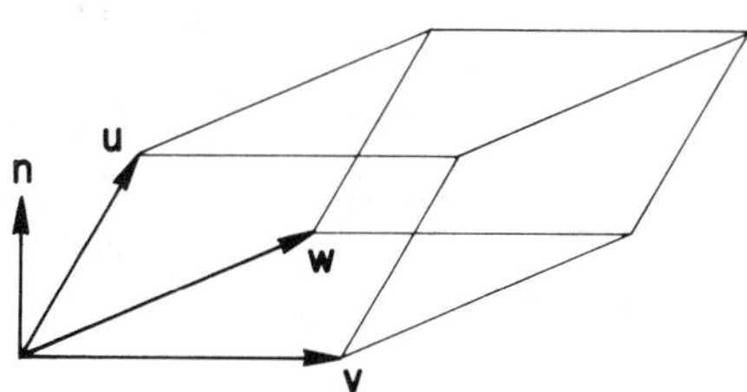

FIGURE 1-4. Parallelepiped of volume $\mathbf{u} \cdot \mathbf{v} \wedge \mathbf{w}$ generated by vectors $\mathbf{u}$, $\mathbf{v}$, and $\mathbf{w}$. Vector $\mathbf{n}$ is a unit vector in the direction of $\mathbf{v} \wedge \mathbf{w}$.

EXERCISE 1–28 Use Eq (1-25) to derive formulas for the unit cell volumes of each of the following.

(a) A monoclinic crystal ($\alpha = \gamma = 90°$).
(b) An orthorhombic crystal ($\alpha = \beta = \gamma = 90°$).
(c) A tetragonal crystal ($a_1 = a_2$, $\alpha = \beta = \gamma = 90°$).
(d) A hexagonal crystal ($a_1 = a_2$, $\alpha = \beta = 90°$, $\gamma = 120°$).
(e) A rhombohedral crystal ($a_1 = a_2 = a_3$, $\alpha = \beta = \gamma$).
(f) A cubic crystal ($a_1 = a_2 = a_3$, $\alpha = \beta = \gamma = 90°$).

EXERCISE 1–29 Calculate the volume of a unit cell with $a_1 = 6.50$, $a_2 = 9.00$, $a_3 = 8.20$ Å, $\alpha = 102.0°$, $\beta = 110.5°$, $\gamma = 96.0°$.

1–10 Reciprocal Basis Vectors

We define three vectors $\mathbf{a}^j$ that have the property that

$$\mathbf{a}_i \cdot \mathbf{a}^j = \delta_i{}^j \tag{1–26}$$

The vectors $\mathbf{a}^j$, which are designated with superscripts, are said to be *reciprocal* basis vectors. In keeping with the convention that subscripts denote covariant quantities and superscripts denote contravariant quantities, the vectors $\mathbf{a}_i$ may be regarded as covariant basis vectors and the reciprocal vectors $\mathbf{a}^i$ as contravariant basis vectors; although the terms covariant and contravariant usually refer to components, it will prove helpful to classify basis vectors in this way also. The full implications of the distinction between covariance and contravariance will become apparent in the development of transformation properties in Chapter 3. Only in cartesian coordinate systems does $\mathbf{a}^i = \mathbf{a}_i$ for $i = 1, 2, 3$.

Equation (1–26) shows, for example, that $\mathbf{a}^1$ is perpendicular to $\mathbf{a}_2$ and $\mathbf{a}_3$ ($\mathbf{a}_2 \cdot \mathbf{a}^1 = 0$, $\mathbf{a}_3 \cdot \mathbf{a}^1 = 0$). Since the vector $\mathbf{a}_2 \Lambda \mathbf{a}_3$ is parallel to $\mathbf{a}^1$, we can write

$$\mathbf{a}^1 = k\mathbf{a}_2 \Lambda \mathbf{a}_3 \tag{1–27}$$

where k is a proportionality constant. Equation (1–26) shows also that $\mathbf{a}_1 \cdot \mathbf{a}^1 = 1$. Hence

$$k = \frac{1}{\mathbf{a}_1 \cdot \mathbf{a}_2 \Lambda \mathbf{a}_3} \tag{1–28}$$

Therefore,

$$\mathbf{a}^1 = \frac{\mathbf{a}_2 \wedge \mathbf{a}_3}{\mathbf{a}_1 \cdot \mathbf{a}_2 \wedge \mathbf{a}_3}, \quad \mathbf{a}^2 = \frac{\mathbf{a}_3 \wedge \mathbf{a}_1}{\mathbf{a}^1 \cdot \mathbf{a}_2 \wedge \mathbf{a}_3},$$
$$\mathbf{a}^3 = \frac{\mathbf{a}_1 \wedge \mathbf{a}_2}{\mathbf{a}_1 \cdot \mathbf{a}_2 \wedge \mathbf{a}_3} \quad (1\text{-}29)$$

If $\mathbf{a}^1$ is in the direction of a right-handed screw that turns $\mathbf{a}_2$ into $\mathbf{a}_3$, $\mathbf{a}_1 \cdot \mathbf{a}_2 \wedge \mathbf{a}_3$ is positive and may be associated with the volume V of the unit cell [by Eq. (1-23)]. For left-handed axes, $\mathbf{a}_1 \cdot \mathbf{a}_2 \wedge \mathbf{a}_3$ is negative and is equal to $-V$. Figure 1-5 shows the axes $\mathbf{a}^1$ and $\mathbf{a}^3$ for a monoclinic unit cell.

The covariant basis vectors are referred to as the basis vectors of the real or direct lattice; the contravariant basis vectors are known to crystallographers as *reciprocal axes.* They are the basis vectors for the reciprocal lattice that is at the heart of many crystallographic calculations. Computations that are complicated by the nonorthogonality of the real lattice axes are sometimes simplified by the introduction of reciprocal basis vectors. The notion of reciprocal space is not the exclusive domain of crystallographers, however, although they are inclined to be possessive about it. P. P. Ewald showed in 1921 that the reciprocal lattice concept could aid in interpreting the diffraction of x rays by crystals, and crystallographers have achieved proficiency in manipulating reciprocal vectors. The idea is more pervasive than just a crystallographer's tool, though, and it contributes a natural simplicity to numerous aspects of nonorthogonal geometry.

Some crystallographers include a wavelength λ in the definition of the reciprocal basis vectors.

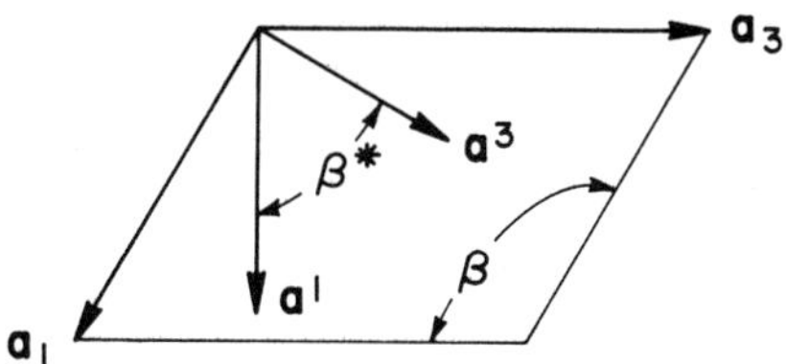

Figure 1-5. Reciprocal axes $\mathbf{a}^1$ and $\mathbf{a}^3$ for a monoclinic crystal. Axes $\mathbf{a}_2$ and $\mathbf{a}^2$ are normal to the paper, pointing inward.

$$\mathbf{a}_i \cdot \mathbf{a}^j = \lambda\, \delta_i^{\,j} \qquad (1\text{-}30)$$

The length of reciprocal lattice vectors then depends upon the details of a particular experiment. Certain diffraction equations are simplified by this definition, but others are made complicated, and such a restricted view of the reciprocal lattice has little to recommend it. Still another definition of reciprocal basis vectors is

$$\mathbf{a}_i \cdot \mathbf{a}^j = 2\pi\delta_i^{\,j} \qquad (1\text{-}31)$$

which is useful in solid-state physics (see, for example, Cornwell, 1969; Kittel, 1971).

EXERCISE 1-30 Calculate the lengths and interaxial angles for the reciprocal axes of the monoclinic crystal of Exercise 1-27.

EXERCISE 1-31 Show that $\mathbf{a}_1 = \mathbf{a}^2 \wedge \mathbf{a}^3/V^*$, $\mathbf{a}_2 = \mathbf{a}^3 \wedge \mathbf{a}^1/V^*$, $\mathbf{a}_3 = \mathbf{a}^1 \wedge \mathbf{a}^2/V^*$, where the axes $\mathbf{a}^i$ form a right-handed set and $V^* = \mathbf{a}^1 \cdot \mathbf{a}^2 \wedge \mathbf{a}^3$ is the volume of the reciprocal cell.

EXERCISE 1-32 Calculate the lengths and interaxial angles for the reciprocal axes of cartesian space.

1-11 Reciprocal Metric Tensor

Any vector in three-dimensional space may be expressed as a linear combination of the three basic vectors. Each of the $\mathbf{a}^i$, therefore, may be written in terms of the $\mathbf{a}_j$. Thus,

$$\mathbf{a}^i = A^{ij}\mathbf{a}_j \qquad (1\text{-}32)$$

where the A^{ij} are coefficients whose values we wish to determine. Taking the scalar product of both sides of Eq. (1-32) with a particular reciprocal basis vector, say $\mathbf{a}^k$, gives

$$\mathbf{a}^i \cdot \mathbf{a}^k = A^{ij}\mathbf{a}_j \cdot \mathbf{a}^k \qquad (1\text{-}33)$$

The term on the left here is the element g^{ik} of the reciprocal (or contravariant) metric tensor, and Eq. (1-26) reduces the right-hand side to A^{ik}. That is,

$$A^{ik} = g^{ik} \tag{1-34}$$

and Eq. (1-32) becomes

$$\mathbf{a}^i = g^{ij}\mathbf{a}_j \tag{1-35}$$

Another handy relationship between the quantities of real and reciprocal space is obtained by taking the scalar product of $\mathbf{a}_k$ with both sides of Eq. (1-35), which gives

$$\delta^i{}_k = g^{ij}g_{jk} \tag{1-36}$$

This equation corresponds to the matrix equation

$$\mathbf{I} = \mathbf{g}^*\mathbf{g} \tag{1-37}$$

where $\mathbf{I}$ is the unit matrix with elements $\delta^i{}_k$,

$$\mathbf{I} = \begin{pmatrix} 1 & 0 & 0 \\ 0 & 1 & 0 \\ 0 & 0 & 1 \end{pmatrix} \tag{1-38}$$

$\mathbf{g}$ is the metric tensor with elements g_{ij} [Eq. (1-17)], and $\mathbf{g}^*$ is the reciprocal metric tensor whose elements g^{ij} are given by

$$g^{ij} = \mathbf{a}^i \cdot \mathbf{a}^j \tag{1-39}$$

Thus, the reciprocal metric tensor is the reciprocal in the matrix algebra sense of the real metric tensor.

We continue our development of reciprocal space by multiplying both sides of Eq. (1-35) by g_{ik} and summing over i:

$$g_{ik}\mathbf{a}^i = g_{ik}g^{ij}\mathbf{a}_j \tag{1-40}$$

By Eq. (1-36), this reduces to

$$g_{ik}\mathbf{a}^i = \mathbf{a}_k \tag{1-41}$$

In matrix notation, Eqs. (1-35) and (1-41) are

$$\mathbf{a}^* = \mathbf{g}^*\mathbf{a} \tag{1-42}$$

$$\mathbf{a} = \mathbf{g}\mathbf{a}^* \tag{1-43}$$

When the reciprocal metric tensor operates on the covariant basis vectors, it produces the contravariant basis vectors. Conversely, the covariant metric tensor converts contravariant basis vectors into covariant basis vectors. The properties that g^{ij} raises suffixes and g_{ij} lowers suffixes are quite general and will be of great assistance to us.

EXERCISE 1-33 A triclinic crystal has $a_1 = 8.00$, $a_2 = 10.00$, $a_3 = 12.00$ Å, $\alpha = 100.0°$, $\beta = 105.0°$, $\gamma = 95.0°$.
(a) Calculate all elements of the metric tensor.
(b) Calculate the reciprocal metric tensor.
(c) Calculate the lengths and angles of the reciprocal lattice.

EXERCISE 1-34 Calculate the lengths and interaxial angles of the reciprocal basis vectors associated with the crystal of Exercise 1-21.

EXERCISE 1-35 Evaluate $g_{rs}g^{rs}$.

EXERCISE 1-36 (a) Evaluate $\delta_r{}^r$. (b) Evaluate $\delta^r{}_s\, \delta^s{}_r$. (c) Evaluate $\delta^r{}_s\, \delta^s{}_t\, \delta^t{}_r$.

EXERCISE 1-37) Let **u** be a unit vector. Define the angle χ_l as the angle between **u** and $\mathbf{a}_l$. In a cartesian coordinate system,

$$\cos^2 \chi_1 + \cos^2 \chi_2 + \cos^2 \chi_3 = 1.$$

Derive the analogous relationship in a general rectilinear system.

1-12 Metric Tensor and Unit Cell Volume

The determinant of the metric tensor [Eq. (1-17)] is

$$g = g_{11}g_{22}g_{33} + g_{12}g_{23}g_{31} + g_{13}g_{32}g_{21} - g_{31}g_{22}g_{13} - g_{21}g_{12}g_{33} - g_{11}g_{23}g_{32} \tag{1-44}$$

Replacing each g_{ij} by its value from Eq. (1-14) in terms of the axial lengths and interaxial angles produces

$$g = (a_1a_2a_3)^2 (1 - \cos^2 \alpha - \cos^2 \beta - \cos^2 \gamma + 2 \cos \alpha \cos \beta \cos \gamma) \tag{1–45}$$

Comparison with Eq. (1–25) reveals that

$$g = V^2 \tag{1–46}$$

Since $\mathbf{g}^*$ is the reciprocal of $\mathbf{g}$, it follows that g^*, the determinant of $\mathbf{g}^*$, is given by

$$g^* = 1/V^2 \tag{1–47}$$

and

$$V^* = 1/V \tag{1–48}$$

Table 1–1 tabulates several useful relationships among the direct and reciprocal unit cell dimensions. A more extensive table of these relationships may be found in Buerger's book (1942). A vector derivation of some of these results has been given by Neustadt, Cagle, and Waser (1968).

EXERCISE 1–38 Calculate the determinants g and g^* of $\mathbf{g}$ and $\mathbf{g}^*$ for the crystal described by Exercises 1–21 and 1–34. Calculate V and V^* directly from Eq. (1–25) and thus verify Eqs. (1–46), (1–47), and (1–48).

EXERCISE 1–39 Calculate g, g^*, and hence V, and V^* for the crystal described by Exercises 1–27 and 1–30.

EXERCISE 1–40 Calculate g, g^*, V, and V^* for the crystal described in Exercise 1–29.

EXERCISE 1–41 Calculate g, g^*, V, and V^* for the crystal described in Exercise 1–33.

EXERCISE 1–42 Show that $a^1 = a_2a_3 \sin \alpha/V$, and write similar relationships for a^2, a^3, a_1, a_2, and a_3.

EXERCISE 1–43 Use the results of Exercise 1–42 and Eq. (1–48) to show that $V = a_1a_2a_3 \sin \alpha^* \sin \beta \sin \gamma$.

TABLE 1-1 SOME USEFUL RELATIONSHIPS AMONG DIRECT AND RECIPROCAL UNIT CELL DIMENSIONS

$$a^1 = a_2a_3 \sin\alpha/V \qquad a_1 = a^2a^3 \sin\alpha^*/V^*$$
$$a^2 = a_3a_1 \sin\beta/V \qquad a_2 = a^3a^1 \sin\beta^*/V^*$$
$$a^3 = a_1a_2 \sin\gamma/V \qquad a_3 = a^1a^2 \sin\gamma^*/V^*$$

$$V = a_1a_2a_3(1 - \cos^2\alpha - \cos^2\beta - \cos^2\gamma + 2\cos\alpha\cos\beta\cos\gamma)^{1/2}$$
$$V^* = a^1a^2a^3(1 - \cos^2\alpha^* - \cos^2\beta^* - \cos^2\gamma^* + 2\cos\alpha^*\cos\beta^*\cos\gamma^*)^{1/2}$$
$$V = a_1a_2a_3 \sin\alpha\sin\beta\sin\gamma^* \qquad V^* = a^1a^2a^3 \sin\alpha^*\sin\beta^*\sin\gamma$$
$$V = a_1a_2a_3 \sin\alpha\sin\beta^*\sin\gamma \qquad V^* = a^1a^2a^3 \sin\alpha^*\sin\beta\sin\gamma^*$$
$$V = a_1a_2a_3 \sin\alpha^*\sin\beta\sin\gamma \qquad V^* = a^1a^2a^3 \sin\alpha\sin\beta^*\sin\gamma^*$$

$$VV^* = 1$$

$$a_1a^1 = 1/(\sin\beta^*\sin\gamma) = 1/(\sin\beta\sin\gamma^*)$$
$$a_2a^2 = 1/(\sin\alpha^*\sin\gamma) = 1/(\sin\alpha\sin\gamma^*)$$
$$a_3a^3 = 1/(\sin\alpha^*\sin\beta) = 1/(\sin\alpha\sin\beta^*)$$

$$\frac{\sin\alpha}{\sin\alpha^*} = \frac{\sin\beta}{\sin\beta^*} = \frac{\sin\gamma}{\sin\gamma^*}$$

$$\cos\alpha = \frac{\cos\beta^*\cos\gamma^* - \cos\alpha^*}{\sin\beta^*\sin\gamma^*}, \qquad \cos\alpha^* = \frac{\cos\beta\cos\gamma - \cos\alpha}{\sin\beta\sin\gamma}$$

$$\cos\beta = \frac{\cos\alpha^*\cos\gamma^* - \cos\beta^*}{\sin\alpha^*\sin\gamma^*}, \qquad \cos\beta^* = \frac{\cos\alpha\cos\gamma - \cos\beta}{\sin\alpha\sin\gamma}$$

$$\cos\gamma = \frac{\cos\alpha^*\cos\beta^* - \cos\gamma^*}{\sin\alpha^*\sin\beta^*}, \qquad \cos\gamma^* = \frac{\cos\alpha\cos\beta - \cos\gamma}{\sin\alpha\sin\beta}$$

EXERCISE 1-44 Given $g_{11} = 25.00$, $g_{12} = -3.17$, $g_{13} = -7.08$, $g^{11} = 42.36 \times 10^{-3}$, $g^{12} = 4.16 \times 10^{-3}$, $g^{22} = 24.54 \times 10^{-3}$, calculate all the lengths and angles of the direct and reciprocal unit cells.

1-13 Covariant Vector Components

Vectors usually are described in terms of their contravariant components, as in Eq. (1-3). The reciprocal basis vectors can serve just as well as a set of linearly independent vectors, so a vector **v** may be written also as

$$\mathbf{v} = v_i \mathbf{a}^i \tag{1-49}$$

Since Eqs. (1-3) and (1-49) describe the same vector,

$$v_i \mathbf{a}^i = v^j \mathbf{a}_j \tag{1-50}$$

Using Eq. (1-41) for $\mathbf{a}_j$,

$$v_i \mathbf{a}^i = v^j g_{ji} \mathbf{a}^i \tag{1-51}$$

from which it follows that

$$v_i = g_{ij} v^j \tag{1-52}$$

Operating on both sides with g^{ki} yields

$$g^{ki} v_i = v^k \tag{1-53}$$

We have thus completed development of a dual vector system, in which any vector can be expressed in terms of covariant basis vectors with contravariant components, or in terms of contravariant basis vectors with covariant components. The metric tensors facilitate conversion from one description to the other.

We shall use the convention of enclosing contravariant components in square brackets, as in $[x^1, x^2, x^3]$, and covariant components in parentheses, as in (u_1, u_2, u_3). In Chapter 2 we shall learn to describe a plane in a crystal by means of its Miller indices (h_1, h_2, h_3), which are the covariant components of the normal to the plane.

The following numerical example will illustrate the application of these results. Consider a vector having $u^1 = 0.500$, $u^2 = 0.000$, $u^3 = 1.750$ in a crystal for which $a_1 = 10.00$, $a_2 = 8.00$, $a_3 = 9.00$ Å, $\alpha = 90.0°$, $\beta = 112.0°$, $\gamma = 90.0°$. We calculate

$$\mathbf{g} = \begin{pmatrix} 100.0 & 0 & -33.7 \\ 0 & 64.0 & 0 \\ -33.7 & 0 & 81.0 \end{pmatrix}$$

$$g = 4.46 \times 10^5, \qquad V = 668$$

$$\mathbf{g}^* = \begin{pmatrix} 0.01163 & 0 & 0.00484 \\ 0 & 0.01563 & 0 \\ 0.00484 & 0 & 0.01436 \end{pmatrix}$$

$$\begin{pmatrix} u_1 \\ u_2 \\ u_3 \end{pmatrix} = \begin{pmatrix} 100.0 & 0 & -33.7 \\ 0 & 64.0 & 0 \\ -33.7 & 0 & 81.0 \end{pmatrix} \begin{pmatrix} 0.500 \\ 0 \\ 1.750 \end{pmatrix} = \begin{pmatrix} -9.00 \\ 0 \\ 124.9 \end{pmatrix}$$

The length of a vector is independent of its mode of description. Thus, the square of the length of a vector is given by the alternative forms

$$\mathbf{u} \cdot \mathbf{u} = u^i g_{ij} u^j = u_i g^{ij} u_j \tag{1-54}$$

EXERCISE 1-45 Calculate the length of the vector **u** in the example above:

(a) By means of $u^i g_{ij} u^j$ applied to the contravariant components [it is convenient to use the matrix form given in Eq. (1-18)].

(b) By means of $u_i g^{ij} u_j$ (again, a matrix formulation facilitates the computation).

EXERCISE 1-46 Convert the vectors of Exercise 1-21 to covariant components and calculate the angle between them by means of $\mathbf{u} \cdot \mathbf{v} = u_i g^{ij} v_j$.

EXERCISE 1-47 Show that Eq. (1-18) is equivalent to $\mathbf{u} \cdot \mathbf{v} = u^i v_i = u_i v^i$

EXERCISE 1-48 Prove that (a) $v_i = \mathbf{v} \cdot \mathbf{a}_i$; (b) $v^i = \mathbf{v} \cdot \mathbf{a}^i$.

1-14 Permutation Tensors

The permutation tensors are defined by

$$\epsilon_{ijk} = \mathbf{a}_i \cdot \mathbf{a}_j \wedge \mathbf{a}_k \tag{1-55}$$

$$\epsilon^{ijk} = \mathbf{a}^i \cdot \mathbf{a}^j \wedge \mathbf{a}^k \tag{1-56}$$

Letting

$$P = \mathbf{a}_1 \cdot \mathbf{a}_2 \wedge \mathbf{a}_3 / V = \mathbf{a}^1 \cdot \mathbf{a}^2 \wedge \mathbf{a}^3 / V^* \tag{1-57}$$

where V is the unit cell volume and V^* is the volume of the reciprocal cell, we obtain $P = +1$ for right-handed axes and -1 for left-handed axes. Then,

$$\epsilon_{ijk} = PV, \qquad \epsilon^{ijk} = PV^* \qquad \text{for even permutations} \tag{1-58}$$

$$\epsilon_{ijk} = -PV, \qquad \epsilon^{ijk} = -PV^* \qquad \text{for odd permutations} \tag{1-59}$$

$$\epsilon_{ijk} = \epsilon^{ijk} = 0 \text{ if } j = i \text{ or } k = i \text{ or } k = j \tag{1-60}$$

where the combinations 123, 231, and 312 are called even permutations and 132, 213, and 321 are odd permutations.

It is convenient also to define the permutation *symbols*

$$\begin{aligned} e_{ijk} = e^{ijk} &= +1 && \text{for even permutations} \\ &= -1 && \text{for odd permutations} \\ &= 0 && \text{if any two indices are equal} \end{aligned} \tag{1-61}$$

The permutation symbol is equivalent to the permutation tensor for a right-handed cartesian system; the general relationships between them are

$$\epsilon_{ijk} = PVe_{ijk}, \qquad \epsilon^{ijk} = PV^*e^{ijk} \tag{1-62}$$

In some respects, it is preferable to use some name other than permutation tensor for ϵ_{ijk} (for example, the epsilon tensor) because permutation is only one of its characteristics; we shall, however, adhere to common usage.

The product $\epsilon_{ijk}\epsilon^{lmn}$ stands for $3^6 = 729$ terms; however, only 36 of these are nonzero. The product $\epsilon_{ijk}\epsilon^{klm}$ has only twelve nonzero combinations; these are

$$\begin{aligned} \epsilon_{123}\epsilon^{312} &= \epsilon_{213}\epsilon^{321} = \epsilon_{132}\epsilon^{213} = \epsilon_{312}\epsilon^{231} \\ &= \epsilon_{231}\epsilon^{123} = \epsilon_{321}\epsilon^{132} = 1 \end{aligned} \tag{1-63}$$

$$\begin{aligned} \epsilon_{123}\epsilon^{321} &= \epsilon_{213}\epsilon^{312} = \epsilon_{132}\epsilon^{231} = \epsilon_{312}\epsilon^{213} \\ &= \epsilon_{231}\epsilon^{132} = \epsilon_{321}\epsilon^{123} = -1 \end{aligned} \tag{1-64}$$

Inspection of Eqs. (1-63) and (1-64) reveals that $\epsilon_{ijk}\epsilon^{kmn}$ equals $+1$ only if $i = m$ and $j = n$, and -1 only if $i = n$ and $j = m$. Hence,

$$\epsilon_{ijk}\epsilon^{kmn} = \delta_i^{\,m}\,\delta_j^{\,n} - \delta_i^{\,n}\,\delta_j^{\,m} \tag{1-65}$$

The inverse **B** of a matrix **A** may be found by replacing each element A^{ij} of **A** by the cofactor of element A^{ji} divided by the determinant A of **A**, where the cofactor of A^{ji} is equal to $(-1)^{j+i}$ times the determinant of the matrix obtained by deleting from **A** the jth row and the ith column. Thus, for a 3×3 case, $B_{11} = (A^{22}A^{33} - A^{32}A^{23})/A$, and $B_{12} = -(A^{12}A^{33} - A^{32}A^{13})/A$, etc. As may be verified by writing out all the terms in detail for the inverse of the metric tensor **g**,

$$g^{ij} = \epsilon^{ikl}\epsilon^{jmn}g_{km}g_{ln}/2 \tag{1-66}$$

and the terms in the inverse of **g*** yield

$$g_{ij} = \epsilon_{ikl}\epsilon_{jmn}g^{km}g^{ln}/2 \tag{1-67}$$

If we multiply both sides of Eq. (1-67) by $\epsilon^{ipq}\epsilon^{jrs}$,

$$g_{ij}\epsilon^{ipq}\epsilon^{jrs} = \epsilon^{ipq}\epsilon^{jrs}\epsilon_{ikl}\epsilon_{jmn}g^{km}g^{ln}/2 \tag{1-68}$$

permute some of the indices,

$$g_{ij}\epsilon^{ipq}\epsilon^{jrs} = \epsilon^{ipq}\epsilon^{jrs}\epsilon_{kli}\epsilon_{mnj}g^{km}g^{ln}/2 \tag{1-69}$$

and apply Eq. (1-65), we obtain

$$g_{ij}\epsilon^{ipq}\epsilon^{jrs} = (\delta_k{}^p\delta_l{}^q - \delta_k{}^q\,\delta_l{}^p) \times (\delta_m{}^r\,\delta_n{}^s - \delta_m{}^s\,\delta_n{}^r)g^{km}g^{ln}/2 \tag{1-70}$$

We get positive contributions on the right when $k = p$, $l = q$, $m = r$, $n = s$, or when $k = q$, $l = p$, $m = s$, $n = r$, and negative contributions when $k = p$, $l = q$, $m = s$, $n = r$, or when $k = q$, $l = p$, $m = r$, $n = s$. Therefore,

$$g_{ij}\epsilon^{ipq}\epsilon^{jrs} = g^{pr}g^{qs} - g^{ps}g^{qr} \tag{1-71}$$

EXERCISE 1-49 Evaluate $\epsilon_{ijk}\epsilon^{ijk}$.

EXERCISE 1-50 By the method of cofactors given above, write out all nine elements of the inverse of a 3×3 matrix, and confirm that the procedure does indeed produce the inverse matrix.

EXERCISE 1-51 Derive a formula analogous to Eq. (1-71) for $g^{ij}\epsilon_{ipq}\epsilon_{jrs}$.

EXERCISE 1-52 Show that the determinant of the 3×3 matrix **A** with elements $A_i{}^j$ can be written $e_{klm}A_1{}^k\,A_2{}^l\,A_3{}^m$, or as $e^{klm}A_k{}^1\,A_l{}^2\,A_m{}^3$.

1-15 Some Vector Relationships

The techniques that have been developed in this chapter make possible the derivation of numerous vector relationships and identities. The following examples illustrate these principles and provide practice with the methods of tensor and vector algebra.

A. Components of Vector Product

Suppose that **u** and **v** are vectors in a three-dimensional space with covariant basis vectors $\mathbf{a}_j$. The vector $\mathbf{u} \wedge \mathbf{v}$ may be written as a linear combination of the $\mathbf{a}_j$,

$$\mathbf{u} \wedge \mathbf{v} = A^j \mathbf{a}_j \tag{1-72}$$

where the A^j are coefficients to be determined. Expressing **u** and **v** in terms of their contravariant components,

$$(u^k \mathbf{a}_k) \wedge (v^l \mathbf{a}_l) = A^j \mathbf{a}_j \tag{1-73}$$

which may be expanded to

$$\begin{aligned}(u^1\mathbf{a}_1 + u^2\mathbf{a}_2 + u^3\mathbf{a}_3) \wedge (v^1\mathbf{a}_1 + v^2\mathbf{a}_2 + v^3\mathbf{a}_3) \\ = A^1\mathbf{a}_1 + A^2\mathbf{a}_2 + A^3\mathbf{a}_3\end{aligned} \tag{1-74}$$

By Eqs. (1-29)

$$\begin{aligned}\mathbf{a}^1(u^2v^3 - u^3v^2) + \mathbf{a}^2(u^3v^1 - u^1v^3) + \\ \mathbf{a}^3(u^1v^2 - u^2v^1) = A^i\mathbf{a}_i/V\end{aligned} \tag{1-75}$$

We convert the basis vectors $\mathbf{a}_i$ to contravariant basis vectors to obtain

$$\begin{aligned}\mathbf{a}^1(u^2v^3 - u^3v^2) + \mathbf{a}^2(u^3v^1 - u^1v^3) + \\ \mathbf{a}^3(u^1v^2 - u^2v^1) = A^i g_{ij}\mathbf{a}^j/V\end{aligned} \tag{1-76}$$

Equating coefficients of $\mathbf{a}^j$ yields the equations

$$\begin{aligned}u^2v^3 - u^3v^2 &= A^i g_{i1}/V \\ u^3v^1 - u^1v^3 &= A^i g_{i2}/V \\ u^1v^2 - u^2v^1 &= A^i g_{i3}/V\end{aligned} \tag{1-77}$$

The permutation tensor enables us to write Eqs. (1-77) concisely as

$$\epsilon_{jkl}u^k v^l = A^i g_{ij} \tag{1-78}$$

The A^i in Eq. (1-78) may be isolated by multiplying both sides by g^{jm}, summing over j, and using Eq. (1-36):

$$A^m = \epsilon_{jkl} g^{jm} u^k v^l \tag{1-79}$$

Hence,

$$\mathbf{u} \wedge \mathbf{v} = \epsilon_{jkl} u^k v^l g^{jm} \mathbf{a}_m \tag{1-80}$$

or

$$\mathbf{u} \wedge \mathbf{v} = \epsilon_{jkl} u^k v^l \mathbf{a}^j \tag{1-81}$$

A special case of this is

$$\mathbf{a}_k \wedge \mathbf{a}_l = \epsilon_{jkl} \mathbf{a}^j \tag{1-82}$$

which should be compared with Eq. (1-29).

EXERCISE 1-53 Verify that Eq. (1-78) is equivalent to Eqs. (1-77) by using Eqs. (1-58) to (1-60) and writing the terms out in full for the three cases $j = 1, j = 2$, and $j = 3$.

EXERCISE 1-54 Show that for the special case of $\mathbf{u} = \mathbf{a}_i$, $\mathbf{v} = \mathbf{a}_j$, Eq. (1-81) reduces to Eq. (1-82), which is identical with Eq. (1-29). (If $\mathbf{u} = \mathbf{a}_i$, $u^k = \delta_i{}^k$, etc.)

EXERCISE 1-55 Show that $\mathbf{u} \cdot \mathbf{v} \wedge \mathbf{w} = \epsilon_{ijk} u^i v^j w^k$.

EXERCISE 1-56 Show that, if two of the vectors of Exercise 1-55 are proportional to each other (for example, $\mathbf{v} = k\mathbf{u}$), then $\mathbf{u} \cdot \mathbf{v} \wedge \mathbf{w} = 0$.

EXERCISE 1-57 Use the result of Exercise 1-55 to calculate the volume of the parallelepiped defined by the vectors [3, 1, 1], [−1, 2.5, 0], and [0, 0, 2] in an orthorhombic crystal for which $a_1 = 7.50$, $a_2 = 9.20$, $a_3 = 7.00$ Å.

EXERCISE 1-58 Show that $\mathbf{u} \wedge \mathbf{v} = \epsilon^{jkl} u_k v_l \mathbf{a}_j$.

B. *Triple Vector Product*

The triple vector product of **u**, **v**, and **w** is **u** Λ (**v** Λ **w**). Since **v** Λ **w** is a vector normal to the plane of **v** and **w**, **u** Λ (**v** Λ **w**) must be normal to **u** and in the plane of **v** and **w**; that is, it is possible to express **u** Λ (**v** Λ **w**) as a linear combination of **v** and **w**. First, by Eq. (1–80),

$$\mathbf{z} = \mathbf{v} \Lambda \mathbf{w} = \epsilon_{ijk} v^j w^k g^{ip} \mathbf{a}_p \tag{1–83}$$

$$\mathbf{u} \Lambda \mathbf{z} = \epsilon_{mnp} u^n z^p g^{mq} \mathbf{a}_q \tag{1–84}$$

Thus, the pth component of z is $\epsilon_{ijk} v^j w^k g^{ip}$. and when we put this into Eq. (1–84) we obtain

$$\mathbf{u} \Lambda (\mathbf{v} \Lambda \mathbf{w}) = \epsilon_{mnp} u^n \epsilon_{ijk} v^j w^k g^{ip} g^{mq} \mathbf{a}_q \tag{1–85}$$

The result of Exercise 1–51 is now applied to $g^{ip}\epsilon_{mnp}\epsilon_{ijk}$ (after permuting the indices to $g^{ip}\epsilon_{ijk}\epsilon_{pmn}$), and we obtain the equations

$$\mathbf{u} \Lambda (\mathbf{v}\Lambda\mathbf{w}) = (g_{jm}g_{kn} - g_{jn}g_{km}) u^n v^j w^k g^{mq} \mathbf{a}_q \tag{1–86}$$

$$\mathbf{u} \Lambda (\mathbf{v}\Lambda\mathbf{w}) = (\delta_j{}^q\, g_{kn} - \delta_k{}^q\, g_{jn}) u^n v^j w^k \mathbf{a}_q \tag{1–87}$$

$$\mathbf{u} \Lambda (\mathbf{v}\Lambda\mathbf{w}) = g_{kn} u^n w^k v^j \mathbf{a}_j - g_{jn} u^n v^j w^k \mathbf{a}_k \tag{1–88}$$

$$\mathbf{u} \Lambda (\mathbf{v}\Lambda\mathbf{w}) = (\mathbf{u} \cdot \mathbf{w})\mathbf{v} - (\mathbf{u} \cdot \mathbf{v})\, \mathbf{w} \tag{1–89}$$

EXERCISE 1–59 Derive a formula for (**u** Λ **v**) Λ **w** as a linear combination of **u** and **v**.

EXERCISE 1–60 Evaluate **u** Λ (**v** Λ **w**) + **v** Λ (**w** Λ **u**) + **w** Λ (**u** Λ **v**).

EXERCISE 1–61 If **u** and **v** are any two vectors, show that $\mathbf{u} = [(\mathbf{u} \cdot \mathbf{v})\mathbf{v} + \mathbf{v} \Lambda (\mathbf{u} \Lambda \mathbf{v})]/v^2$, which resolves **u** into components parallel and perpendicular to **v**.

C. *Evaluation of* (**u** Λ **v**) • (**w** Λ **z**)

Evaluation of (**u** Λ **v**) • (**w** Λ **z**) is simplified greatly by judicious selection of contravariant and covariant components (see Shmueli, 1974). From Eqs. (1–80) and 1–81),

$$(\mathbf{u} \wedge \mathbf{v}) \cdot (\mathbf{w} \wedge \mathbf{z}) = (\epsilon_{jkl} u^k v^l g^{jm} \mathbf{a}_m) \cdot (\epsilon_{pqr} w^q z^r \mathbf{a}^p) \tag{1-90}$$

$$(\mathbf{u} \wedge \mathbf{v}) \cdot (\mathbf{w} \wedge \mathbf{z}) = \epsilon_{jkl} \epsilon_{pqr} u^k v^l w^q z^r g^{jm} \delta_m{}^p \tag{1-91}$$

$$(\mathbf{u} \wedge \mathbf{v}) \cdot (\mathbf{w} \wedge \mathbf{z}) = g^{jp} \epsilon_{jkl} \epsilon_{pqr} u^k v^l w^q z^r \tag{1-92}$$

By Exercise 1–51,

$$(\mathbf{u} \wedge \mathbf{v}) \cdot (\mathbf{w} \wedge \mathbf{z}) = (g_{kq} g_{lr} - g_{kr} g_{lq}) u^k v^l w^q z^r \tag{1-93}$$

The lowering property of the metric tensor converts this to

$$(\mathbf{u} \wedge \mathbf{v}) \cdot (\mathbf{w} \wedge \mathbf{z}) = u_q v_r w^q z^r - u_r v_q w^q z^r \tag{1-94}$$

and by Exercise 1–47 we have finally

$$(\mathbf{u} \wedge \mathbf{v}) \cdot (\mathbf{w} \wedge \mathbf{z}) = (\mathbf{u} \cdot \mathbf{w})(\mathbf{v} \cdot \mathbf{z}) - (\mathbf{u} \cdot \mathbf{z})(\mathbf{v} \cdot \mathbf{w}) \tag{1-95}$$

D. Evaluation of $(\mathbf{u} \wedge \mathbf{v}) \wedge (\mathbf{w} \wedge \mathbf{z})$

While evaluation of $(\mathbf{u} \wedge \mathbf{v}) \wedge (\mathbf{w} \wedge \mathbf{z})$ might be offered as an exercise for sharpening skills, we shall present the essential steps of the process. Applying Eq. (1–80) to both $\mathbf{u} \wedge \mathbf{v}$ and $\mathbf{w} \wedge \mathbf{z}$,

$$(\mathbf{u} \wedge \mathbf{v}) \wedge (\mathbf{w} \wedge \mathbf{z}) = (\epsilon_{jkl} u^k v^l g^{jm} \mathbf{a}_m) \wedge (\epsilon_{pqr} w^q z^r g^{ps} \mathbf{a}_s) \tag{1-96}$$

By Eq. (1–82) this is

$$(\mathbf{u} \wedge \mathbf{v}) \wedge (\mathbf{w} \wedge \mathbf{z}) = \epsilon_{jkl} \epsilon_{pqr} u^k v^l w^q z^r g^{jm} g^{ps} \epsilon_{tms} \mathbf{a}^t \tag{1-97}$$

which is reduced by Exercise 1–51 to

$$(\mathbf{u} \wedge \mathbf{v}) \wedge (\mathbf{w} \wedge \mathbf{z}) = (g_{ks} g_{lt} - g_{kt} g_{sl}) u^k v^l w^q z^r g^{ps} \epsilon_{pqr} \mathbf{a}^t \tag{1-98}$$

Application of Eqs. (1–36) and (1–52) produces

$$(\mathbf{u} \wedge \mathbf{v}) \wedge (\mathbf{w} \wedge \mathbf{z}) = (u^p v_t w^q z^r - u_t v^p w^q z^r) \epsilon_{pqr} \mathbf{a}^t \tag{1-99}$$

By Eq. (1–50) we can replace $v_t \mathbf{a}^t$ by $v^t \mathbf{a}_t$ and $u_t \mathbf{a}^t$ by $u^t \mathbf{a}_t$, so

$$(\mathbf{u} \wedge \mathbf{v}) \wedge (\mathbf{w} \wedge \mathbf{z}) = \epsilon_{pqr} (u^p w^q z^r v^t \mathbf{a}_t - v^p w^q z^r u^t \mathbf{a}_t) \tag{1-100}$$

and finally by Exercise 1-55

$$(\mathbf{u} \wedge \mathbf{v}) \wedge (\mathbf{w} \wedge \mathbf{z}) = (\mathbf{u} \cdot \mathbf{w} \wedge \mathbf{z})\mathbf{v} - (\mathbf{v} \cdot \mathbf{w} \wedge \mathbf{z})\mathbf{u} \tag{1-101}$$

An alternative pathway from Eq. (1-97) applies the result of Exercise 1-51 to $g^{ps}\epsilon_{pqr}\epsilon_{stm}$ and obtains the equivalent fornula

$$(\mathbf{u} \wedge \mathbf{v}) \wedge (\mathbf{w} \wedge \mathbf{z}) = (\mathbf{u} \cdot \mathbf{v} \wedge \mathbf{z})\mathbf{w} - (\mathbf{u} \cdot \mathbf{v} \wedge \mathbf{w})\mathbf{z} \tag{1-102}$$

Since $(\mathbf{u} \wedge \mathbf{v}) \wedge (\mathbf{w} \wedge \mathbf{z})$ lies both in the plane of **u** and **v** and in the plane of **w** and **z**, it defines the intersection of these planes.

EXERCISE 1-62 A crystal has $a_1 = 8.80$, $a_2 = 9.50$, $a_3 = 7.00$ Å, $\alpha = 102.5°$, $\beta = 97.0°$, $\gamma = 108.5°$. Atom A is at [0.200, 0.050, 0.005], atom B is at [0.100, −0.100, 0.100], and atom C is at [0.000, 0.000, 0.200], where the coordinates are all contravariant components. Find the contravariant components of a unit vector normal to the plane of *A*, *B*, and *C*.

EXERCISE 1-63 A compound with the formula $C_4H_6OSO_2$ crystallizes in the orthorhombic system with $a_1 = 8.475$, $a_2 = 10.742$, $a_3 = 5.899$ Å. The following atomic coordinates have been reported.

S	0.134	0.112	0.162
O(1)	0.301	0.085	0.142
O(2)	0.084	0.186	0.353
O(3)	−0.200	0.093	0.004
C(1)	−0.097	−0.012	−0.032
C(2)	−0.077	0.101	−0.163

(a) Calculate the lengths of the S-O(1), S-O(2), C(1)-O(3), and C(2)-O(3) bonds.

(b) Calculate the O(1)-S-O(2) bond angle and the C(1)-O(3)−C(2) bond angle.

(c) Calculate the vector defined by the intersection of the O(1)-S-O(2) plane with the C(1)-O(3)-C(2) plane. Use both Eq. (1-101) and Eq. (1-102) and verify that they give the same vector.

(d) Calculate the angle between the O(1)-S-O(2) plane and the C(1)-O(3)-C(2) plane (this is the supplement of the angle between the normals to the planes).

CHAPTER 2

Lines and Planes

2–1 Introduction

The location of a point in three-dimensional space is specified by three coordinates relative to some choice of basis vectors. If a functional relationship

$$f(x^1, x^2, x^3) = 0 \tag{2–1}$$

exists among the coordinates of each point, the set of points is confined to a surface, and the geometry of the set is essentially two-dimensional. In general, the surface is curved, but if the relationship is linear (only the first powers of coordinates present), the points lie on a plane. A plane, therefore, is a set of points whose coordinates are related by one linear relationship.

If the coordinates of the points of a set are related by two functional relationships, the points lie on a curve, and the set is one-dimensional. If these two relationships are linear, the points lie on a straight line.

This chapter will be devoted to ways of specifying lines and planes and to methods of calculating some of their properties. Not only do these procedures have numerous applications in themselves, but the results obtained will be essential to the development of the material in the chapters that follow.

2-2 Specification of a Line by Two Points

If U and V, with coordinates u^i and v^i, are any two points on a line, the line is parallel to $\mathbf{u} - \mathbf{v}$ (or $\mathbf{v} - \mathbf{u}$, see Fig. 2-1). If now X, with coordinates x^i, is also on the line, the vector $\mathbf{x} - \mathbf{v}$ (where x has components x^i) is parallel to $\mathbf{u} - \mathbf{v}$. An expression of this parallelism of vectors is

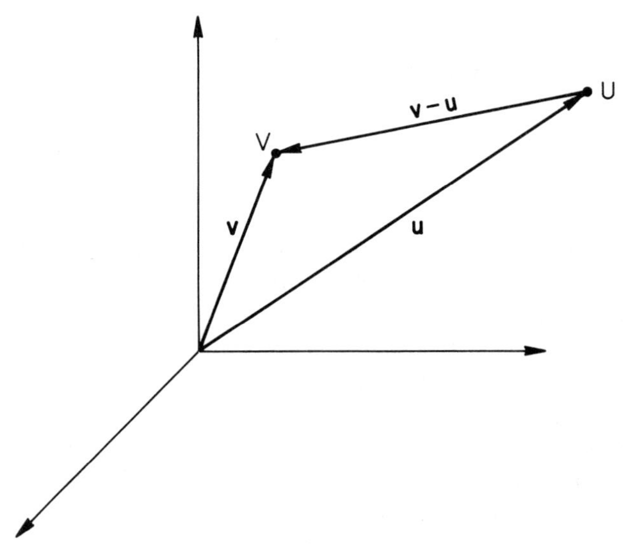

FIGURE 2-1. Two points determine a line.

$$(\mathbf{x} - \mathbf{v}) \wedge (\mathbf{u} - \mathbf{v}) = 0 \qquad (2\text{–}2)$$

which may be expanded to

$$(\mathbf{x} \wedge \mathbf{u}) + (\mathbf{u} \wedge \mathbf{v}) + (\mathbf{v} \wedge \mathbf{x}) = 0 \qquad (2\text{–}3)$$

or, in suffix notation,

$$\epsilon_{ijk}(x^j u^k + u^j v^k + v^j x^k)\mathbf{a}^i = 0 \qquad (2\text{–}4)$$

A vector can be equal to zero only if each component is zero, so for each i,

$$\epsilon_{ijk}(x^j u^k + u^j v^k + v^j x^k) = 0 \qquad (2\text{–}5)$$

There is one redundancy in the three equations represented by Eq. (2–5), so there are only two independent relationships among the x^i.

Exercise 2–1 (a) Show that Eq. (2–5) is equivalent to the matrix equation

$$\begin{pmatrix} 0 & (u^3 - v^3) & -(u^2 - v^2) \\ -(u^3 - v^3) & 0 & (u^1 - v^1) \\ (u^2 - v^2) & -(u^1 - v^1) & 0 \end{pmatrix} \begin{pmatrix} x^1 \\ x^2 \\ x^3 \end{pmatrix} = \begin{pmatrix} u^3v^2 - u^2v^3 \\ u^1v^3 - u^3v^1 \\ u^2v^1 - u^1v^2 \end{pmatrix}$$

(b) Show that the matrix in (a) is singular (that is, its determinant is zero), which amounts to proving that the x^i are not linearly independent.

(c) Show that none of the 2×2 matrices produced by striking out row i and column i of the matrix in (a) is singular. This verifies that the number of independent coordinates is two.

Exercise 2–2 Find the equations relating the coordinates of points along the line connecting the point [0.220, 0.100, 0] to the point [0.150, −0.050, 0.320].

2–3 Specification of a Line by Its Direction and One Point

Consider a line that goes through point U and is parallel to vector $\mathbf{v}$ (see Fig. 2–2). If $\mathbf{x}$ is a vector to any point X on the line, $\mathbf{x} - \mathbf{u}$ is required to be parallel to $\mathbf{v}$. Hence,

$$(\mathbf{x} - \mathbf{u}) \wedge \mathbf{v} = 0 \qquad (2\text{–}6)$$

$$\epsilon_{ijk}(x^j - u^j)v^k\mathbf{a}^i = 0 \qquad (2\text{–}7)$$

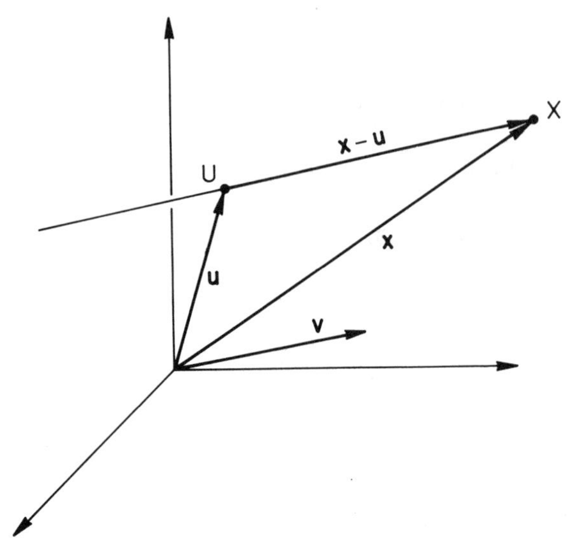

FIGURE 2–2. Line UX determined by point U and direction $\mathbf{v}$.

For each i,

$$\epsilon_{ijk}(x^j - u^j)v^k = 0 \tag{2–8}$$

As was the case with Eq. (2–5), the three equations denoted by Eq. (2–8) correspond to only two independent relationships among the x^i.

EXERCISE 2–3 Write Eq. (2–8) in matrix form similar to that of Exercise 2–1. Show that this matrix is singluar, but that none of the 2 × 2 submatrices produced by deleting a row and the corresponding column is singluar.

EXERCISE 2–4 Find the relationships among the coordinates x^i of points on a line through the point [0.125, −0.200, 0.065] parallel to the vector [−0.100, 0, 0.160].

2–4 Line through a Point and Perpendicular to Two Vectors

The coordinates x^i of points along a line through a point U and perpendicular to vectors $\mathbf{v}$ and $\mathbf{w}$ are given by

$$(\mathbf{x} - \mathbf{u}) \Lambda (\mathbf{v} \Lambda \mathbf{w}) = 0 \tag{2-9}$$

where **u** is the vector from the origin to U (see Fig. 2-3). By Eq. (1-89),

$$[(\mathbf{x} - \mathbf{u}) \cdot \mathbf{w}]\mathbf{v} - [(\mathbf{x} - \mathbf{u}) \cdot \mathbf{v}]\mathbf{w} = 0 \tag{2-10}$$

Each of the terms in square brackets must vanish, so the coordinates x^i must obey the two relationships

$$(x^i - u^i)w^j g_{ij} = 0, \qquad (x^i - u^i)v^j g_{ij} = 0 \tag{2-11}$$

The matrix form of these equations is

$$\tilde{\mathbf{x}}\mathbf{g}\mathbf{w} = \tilde{\mathbf{u}}\mathbf{g}\mathbf{w}, \qquad \tilde{\mathbf{x}}\mathbf{g}\mathbf{v} = \tilde{\mathbf{u}}\mathbf{g}\mathbf{v} \tag{2-12}$$

EXERCISE 2-5 Find the equations relating the coordinates of points on the line perpendicular to vectors $\mathbf{v} = [0.100, 0, -0.100]$ and $\mathbf{w} = [0.060, 0.200, 0.150]$ going through the point [0.400, 0, 0] in a coordinate system with $a_1 = 7.00$, $a_2 = 6.00$, $a_3 = 8.00$ Å, $\alpha = 90.00°$, $\beta = 110.00°$, $\gamma = 100.00°$.

2-5 Projection of One Vector onto Another

The projection of vector **v** onto vector **w** is a vector in the direction of **w** of length $\mathbf{v} \cdot \mathbf{w}/w$. Denoting this vector by $\mathbf{P}(\mathbf{v}/\mathbf{w})$,

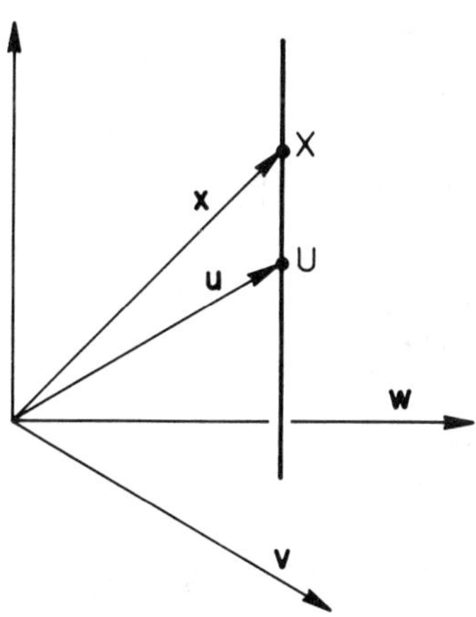

FIGURE 2-3. Line through U normal to **v** and **w**.

$$\mathbf{P}(\mathbf{v}/\mathbf{w}) = (\mathbf{v} \cdot \mathbf{w})\mathbf{w} \,/\, (w)^2 \tag{2-13}$$

EXERCISE 2-6 Find the projection of vector [0.100, 0, −0.100] of Exercise 2-5 onto [0.060, 0.200, 0.150]

EXERCISE 2-7 Show that the projection of vector $\mathbf{v}$ onto axis $\mathbf{a}_I$ is

$$v_I \mathbf{a}_I/(\mathrm{a}_I)^2 \qquad \text{(no summation over } I\text{)}$$

2-6 Plane Determined by Three Points

Three points that do not lie in a straight line determine a plane. If the points are U, V, and W, lying at the ends of vectors $\mathbf{u}$, $\mathbf{v}$, and $\mathbf{w}$, respectively, the vectors $\mathbf{u} - \mathbf{v}$, $\mathbf{v} - \mathbf{w}$, and $\mathbf{w} - \mathbf{u}$ lie in (or at least parallel to) the plane (see Fig. 2-4). The vector

$$\mathbf{z} = (\mathbf{u} - \mathbf{v}) \wedge (\mathbf{v} - \mathbf{w}) \tag{2-14}$$

is normal to this plane; expansion of this equation produces

$$\mathbf{z} = (\mathbf{u} \wedge \mathbf{v}) + (\mathbf{v} \wedge \mathbf{w}) + (\mathbf{w} \wedge \mathbf{u}) \tag{2-15}$$

which by Eq. (1-81) is

$$\mathbf{z} = \epsilon_{ijk}(u^j v^k + v^j w^k + w^j u^k)\mathbf{a}^i \tag{2-16}$$

The magnitude of $\mathbf{z}$ is equal to the area of the parallelogram defined by any two of the vectors $\mathbf{u} - \mathbf{v}$, $\mathbf{v} - \mathbf{w}$, and $\mathbf{w} - \mathbf{u}$; hence, the area of the triangle depicted in Fig. 2-4 is $z/2$.

If $\mathbf{x}$ is a vector from the origin to any point in the plane, $\mathbf{x} - \mathbf{u}$ is in the plane, and

$$(\mathbf{x} - \mathbf{u}) \cdot \mathbf{z} = 0 \tag{2-17}$$

From Eq. (2-15),

$$\mathbf{u} \cdot \mathbf{z} = \mathbf{u} \cdot \mathbf{v} \wedge \mathbf{w} \tag{2-18}$$

and replacing $\mathbf{u} \cdot \mathbf{z}$ by $\mathbf{x} \cdot \mathbf{z}$ [from Eq. (2-17)] and using the covariant components of $\mathbf{z}$ [from Eq. (2-16)] gives

$$\epsilon_{ijk} x^i (u^j v^k + v^j w^k + w^j u^k) = \epsilon_{ijk} u^i v^j w^k \tag{2-19}$$

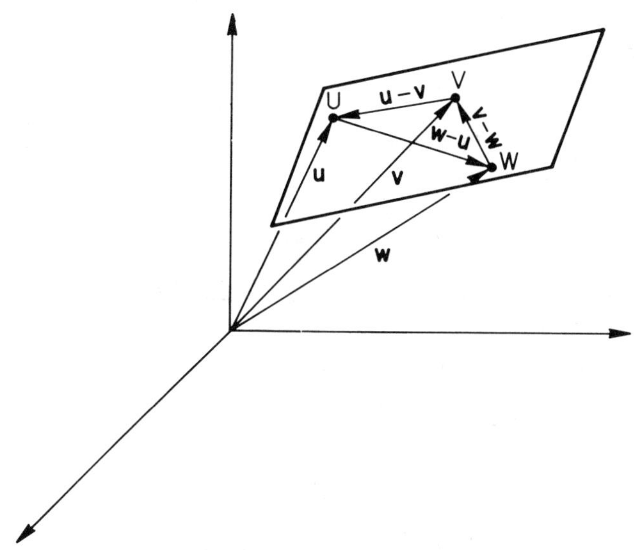

FIGURE 2-4. The plane determined by points U, V and W.

EXERCISE 2-8 Show that Eq. (2-19) is equivalent to the determinantal equation

$$\begin{vmatrix} x^1 & x^2 & x^3 \\ v^1 & v^2 & v^3 \\ w^1 & w^2 & w^3 \end{vmatrix} + \begin{vmatrix} u^1 & u^2 & u^3 \\ x^1 & x^2 & x^3 \\ w^1 & w^2 & w^3 \end{vmatrix} + \begin{vmatrix} u^1 & u^2 & u^3 \\ v^1 & v^2 & v^3 \\ x^1 & x^2 & x^3 \end{vmatrix} = \begin{vmatrix} u^1 & u^2 & u^3 \\ v^1 & v^2 & v^3 \\ w^1 & w^2 & w^3 \end{vmatrix}$$

EXERCISE 2-9 Find the equation obeyed by the coordinates of points lying on the plane that passes through the points [0, 0.520, 0.160], [0.250, −0.250, 0.100], [0.100, 0, 0.100].

2-7 Plane Normal to a Vector and through a Given Point

Points on a plane that is perpendicular to vector **v** and that goes through a point U are given by

$$(\mathbf{x} - \mathbf{u}) \cdot \mathbf{v} = 0 \qquad (2\text{-}20)$$

$$(x^i - u^i)g_{ij}v^j = 0 \qquad (2\text{-}21)$$

where **u** is the vector from the origin to U. The matrix form of this equation is given by Eq. (2–12).

EXERCISE 2–10 Find the equation satisfied by points on the plane that is normal to the vector [0, 0.500, 0.500] and that passes through the point [0.150, −0.100, 0] in the coordinate system of Exercise 2–5.

2–8 Intercepts and Indices

Suppose that a plane intersects the basis vectors at points U, V, and W. That is,

$$\begin{aligned} \mathbf{u} &= [u^1, 0, 0] \\ \mathbf{v} &= [0, v^2, 0] \\ \mathbf{w} &= [0, 0, w^3] \end{aligned} \qquad (2\text{–}22)$$

are three points on the plane. By Eq. (2–19) the coordinates x^i of a point X on the plane satisfy

$$x^1/u^1 + x^2/v^2 + x^3/w^3 = 1 \qquad (2\text{–}23)$$

If the intercepts are replaced by their reciprocals, with

$$h_1 = 1/u^1, \qquad h_2 = 1/v^2, \qquad h_3 = 1/w^3 \qquad (2\text{–}24)$$

Eq. (2–23) becomes

$$x^i h_i = 1 \qquad (2\text{–}25)$$

The quantities h_i are called the *Miller indices* of the plane. If the plane is parallel to one of the basis vectors, the intercept on that axis is taken to be infinity, and the corresponding index is zero. (The use of covariant indices on the h_i will be justified in Section 3–6.) A plane whose indices are h_1, h_2, h_3 will be denoted by (h_1, h_2, h_3).

The numbers h_i define a vector **h**,

$$\mathbf{h} = h_i\mathbf{a}^i \tag{2-26}$$

If

$$\mathbf{x} = x^i\mathbf{a}_i \tag{2-27}$$

is a vector to any point on the plane, Eq. (2-25) is equivalent to

$$\mathbf{h} \cdot \mathbf{x} = 1 \tag{2-28}$$

If **u**, **v**, and **w** are given by Eq. (2-22), vectors $\mathbf{u} - \mathbf{v}$, $\mathbf{v} - \mathbf{w}$, and $\mathbf{w} - \mathbf{u}$ lie in the plane whose intercepts are U, V, and W. From Eqs. (2-22) and (2-24),

$$\mathbf{h} \cdot (\mathbf{u} - \mathbf{v}) = 0 \tag{2-29}$$

Similarly, $\mathbf{h} \cdot (\mathbf{v} - \mathbf{w})$ and $\mathbf{h} \cdot (\mathbf{w} - \mathbf{u})$ are zero. This demonstrates the tremendously important property that **h** is normal to the plane it describes. When the plane is a lattice plane in a crystal, **h** is referred to as a reciprocal lattice vector. The representation of planes by their normals permits applying the rather simple techniques of vector algebra to what otherwise could be rather formidable calculations.

The covariant indices of a plane may be obtained now in terms of the contravariant coordinates of three points that lie on the plane. If these points are U, V, and W, vector **h** must be parallel to **z** of Eq. (2-16); that is,

$$\mathbf{h} = C\mathbf{z} \tag{2-30}$$

where C is a scale factor. Thus,

$$h_i = C\epsilon_{ijk}(u^jv^k + v^jw^k + w^ju^k) \tag{2-31}$$

By Eq. (2-28), the scalar product of **h** with **u** is 1, and taking the scalar product of **u** with $C\mathbf{z}$ of Eq. (2-15) shows that

$$C = 1/\mathbf{u} \cdot \mathbf{v} \wedge \mathbf{w} \tag{2-32}$$

Therefore,

$$h_i = \epsilon_{ijk}(u^j v^k + v^j w^k + w^j u^k)/\epsilon_{lmn}u^l v^m w^n \tag{2-33}$$

EXERCISE 2–11 Write the indices of the planes whose intercepts are (a) [1/2, 1/3, ∞]; (b) [2, 1, 1/5]; (c) [1/3, 1/6, 1/2]; (d) [2/3, 1/3, 1].

EXERCISE 2–12 Write out the covariant components of vector **z** of Eq. (2–14) for the special case of Eq. (2–22); then determine the components of **z** Λ **h**, where h_i is given by Eq. (2–24), and thus confirm that **h** is normal to the plane whose intercepts are $1/h_i$. (Note: Exercise 1–58 gives **u** Λ **v** in terms of covariant components of **u** and **v**.)

EXERCISE 2–13 Find the indices and the intercepts of the plane of Exercise 2-9.

EXERCISE 2–14 Find the equation of the plane that passes through the points [3, 2, 1], [1, 4, −2], and [−2, 0, 1]. Find the intercepts and indices of the plane.

EXERCISE 2–15 Find the indices of the plane that goes through point U and is perpendicular to vector **v**.

EXERCISE 2–16 Find the indices and intercepts of the plane of Exercise 2–10 (a) by using the result of Exercise 2–10, (b) by using the result of Exercise 2–15.

2–9 Length of Normal to Plane

Let **h** be the normal to a plane with indices h_i, and let C be a scale factor such that $C\mathbf{h}$ terminates on the plane. If **x** is a vector to any point on the plane, vector $\mathbf{x} - C\mathbf{h}$ is parallel to the plane. Hence,

$$(\mathbf{x} - C\mathbf{h}) \cdot \mathbf{h} = 0 \tag{2-34}$$

Using Eq. (2–28)

$$C = 1/(h)^2 \tag{2-35}$$

Thus, the vector $\mathbf{h}/(h)^2$ terminates on the plane, and the distance from the origin to the plane is $1/h$.

EXERCISE 2–17 Find the distance from the origin to the plane of Exercise 2–10 (see Exercise 2–16).

EXERCISE 2–18 Calculate the distance from the origin to the plane of Exercise 2-9. (Use the cell dimensions of Exercise 2–5).

2-10 Distance from a Point to a Plane

The distance l from a point U to a plane with indices h_i is equal to $1/h$ minus the length of **P(u/h)**, where **P(u/h)** is the projection of **u** onto **h** (see Fig. 2-5), and **u** is the vector from the origin to U.

From Eq. (2-13)

$$l = |(1 - \mathbf{u} \cdot \mathbf{h})|/h \tag{2-36}$$

If the term $1 - \mathbf{u} \cdot \mathbf{h}$ is positive, the point is on the same side of the plane as the origin, and if $1 - \mathbf{u} \cdot \mathbf{h}$ is negative, the point is beyond the plane from the origin.

EXERCISE 2-19 Calculate the distance from the point [0.300, 0.100, 0.300] to the plane of Exercise 2-18.

2-11 Lattice Planes

If h_1, h_2, and h_3 are relatively prime integers (greatest common divisor equal to 1), an important result from the theory of numbers is

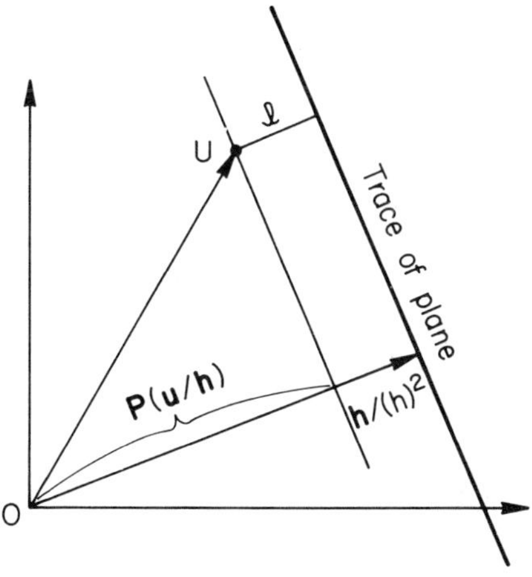

FIGURE 2-5. Distance l from point U to a plane with normal **h**.

that the Diophantine equation, Eq. (2–25), has integer solutions x^i. Since points with integral contravariant coodinates are lattice points, such a plane passes through an infinite number of lattice points. Planes with this property are called *lattice planes*. The indistinguishability of lattice points from each other implies that planes identical in orientation and in environment (that is, with identical atoms distributed identically with respect to the plane) must pass through every lattice point. A plane whose indices are relatively prime integers is thus one of an infinite set of identical, parallel, equidistant planes. The distance between these planes is

$$d = 1/|h_1\mathbf{a}^1 + h_2\mathbf{a}^2 + h_3\mathbf{a}^3| = 1/h \tag{2–37}$$

If the indices are rational numbers, but not integers, Eq. (2–25) can be converted to an integer equation by multiplying through by the least common denominator c to obtain

$$cx^ih_i = c \tag{2–38}$$

If now the integers ch_1, ch_2, ch_3 are relatively prime, Eq. (2–38) is another Diophantine equation with integer solutions, and it also describes a plane that passes through lattice points. The equation

$$cx^ih_i = 1 \tag{2–39}$$

represents a plane parallel to that of Eq. (2–38), so again we have a set of identical, parallel, equidistant lattice planes. The 1 on the right of Eq. (2–39) defines the lattice plane closest to that which passes through the origin; thus, the distance between adjacent planes of this set is $1/ch$.

If the integers h_i [or ch_i in the case of Eqs. (2–38) and (2–39)] are all divisible by an integer other than 1, the plane $(h_1h_2h_3)$ will not pass through lattice points [except for some special cases in nonprimitive unit cells; for example, (200) is a lattice plane in a B-centered, C-centered, or body-centered cell]. However, any plane with rational intercepts is parallel to a set of lattice planes, and it is customary for crystallographers to treat all planes with integral indices as lattice planes. Although (400), for example, would pass through lattice points only for a very unusual choice of a nonprimitive cell, crystallographers are likely to refer to the (400) x-ray reflection rather than to the fourth order of the (100) reflection. The set of planes designated as (400) includes the (100) planes and the intervening planes such that the interplanar spacing is one-quarter that of (100). While these planes are parallel and equidistant, they are not identical in the sense of passing through identical atomic or molecular environments.

Denoting the interplanar spacing by d [as in Eq. (2–37)],

$$1/(d)^2 = h_i h_j g^{ij} \tag{2–40}$$

The quantity d is the spacing determined by means of the Bragg equation,

$$\lambda = 2d \sin \theta \tag{2–41}$$

which will be examined in detail in Chapter 6.

EXERCISE 2–20 Verify that the plane (321) goes through the lattice points given by $x^1 = (3 + 2n)(3 + m)$, $x^2 = (-4 - 3n)(3 + m)$, $x^3 = -2 - m$, where m and n are any integers.

EXERCISE 2–21 Write Eq. (2–38) for the plane (2/5, 4/5, 6/5). Does this plane pass through the lattice points of a primitive unit cell?

EXERCISE 2–22 Calculate the interplanar spacing of (210) and of (111) in the crystal of Exercise 2–5.

EXERCISE 2–23 Derive formulas for the interplanar spacings of planes in the following systems. Express the results in terms of the direct lattice dimensions. (a) Monoclinic. (b) Orthorhombic. (c) Tetragonal. (d) Hexagonal. (e) Rhombohedral. (f) Cubic.

2–12 Angle between Planes

The angle between two planes is the supplement of the angle between the normals to the planes (see Fig. 2–6). The problem of determining the angle between two planes thus devolves to that of finding the angle between two vectors. If $\mathbf{h}$ and $\mathbf{h}'$ are the normals to the planes

$$\mathbf{h} \cdot \mathbf{h}' = hh' \cos(\pi - \phi) \tag{2–42}$$

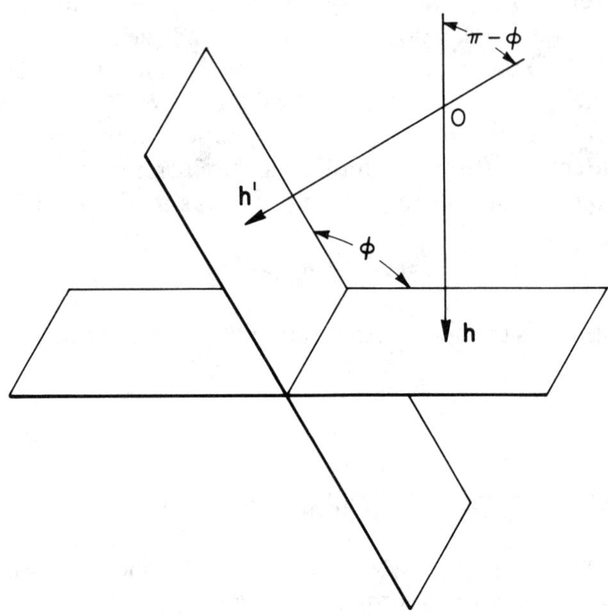

FIGURE 2–6. Angle between two planes

$$\cos(\pi - \phi) = h_i h'_j g^{ij} d d' \tag{2–43}$$

where ϕ is the angle between the planes, and d and d' are the respective interplanar spacings.

EXERCISE 2–24 Calculate the angles between the plane of Exercise 2–10 (see also Exercises 2–5, 2–16, and 2–17), and (a) (001), (b) (100), (c) (010), (d) (210), (e) (021).

EXERCISE 2–25 Calculate the angle between the (210) and (111) planes in the crystal of Exercise 2–5.

EXERCISE 2–26 Calculate the angles between the following pairs of planes in a cubic crystal: (a) (111) and ($11\bar{1}$) (Note: A bar over a number indicates a minus sign, $\bar{1} = -1$.) (b) (111) and (110). (c) (111) and ($1\bar{1}0$). (d) (111) and (100). (e) (210) and (111).

2–13 Projection of a Vector onto a Plane

The projection of a vector **u** onto a plane described by reciprocal vector **h** normal to the plane is the vector

$$\mathbf{P}(\mathbf{u}\,\|\,\mathbf{h}) = \mathbf{h}\;\Lambda\;(\mathbf{u}\;\Lambda\;\mathbf{h})/(h)^2 \tag{2–44}$$

The notation is similar to that used in Section 2–5 for the projection of one vector onto another, except that in this case the projection is parallel to vector **h** onto the plane perpendicular to **h**. By Eq. (1–89)

$$\mathbf{P}(\mathbf{u}\,\|\,\mathbf{h}) = \mathbf{u} - (\mathbf{h}\cdot\mathbf{u})\mathbf{h}/(h)^2 \tag{2–45}$$

The contravariant components of this vector are

$$\mathrm{P}^i(\mathbf{u}\,\|\,\mathbf{h}) = u^i - (h_j u^j)g^{ik}h_k/(h)^2 \tag{2–46}$$

and the covariant components are

$$\mathrm{P}_i(\mathbf{u}\,\|\,\mathbf{h}) = u_i - (h_j u^j)h_i/(h)^2 \tag{2–47}$$

EXERCISE 2–27 Derive Eq. (2–45) from Eq. (2–13) by finding the projection of **u** onto the vector **h** Λ (**u** Λ **h**).

EXERCISE 2–28 Find the projections of vectors **v** and **w** of Exercise 2–5 onto the (201) plane, calculate the angle between the projected vectors, and compare with the angle between **v** and **w**.

EXERCISE 2–29 A monoclinic crystal has a_1 = 7.68, a_2 = 9.04, a_3 = 8.51 Å, β = 104.3°. Calculate the covariant components of the projection of the reciprocal lattice vector ($\bar{1}$32) onto the (001) plane.

EXERCISE 2–30 Calculate the covariant components of the projection of the reciprocal lattice vector ($\bar{1}$32) of Exercise 2–29 onto the plane normal to $\mathbf{a}_3$.

EXERCISE 2–31 Calculate the covariant components of the projection of reciprocal lattice vector (210) in the coordinate system of Exercises 2–29 and 2–30 onto (a) the (001) plane, (b) the plane normal to $\mathbf{a}_3$.

CHAPTER 3

Transformations

3–1 Introduction

A transformation consists of rules by which a new set of basis vectors is obtained from an original set of basis vectors. The transformations of the greatest utility and interest are those that are unique and single-valued; starting with a set $\mathbf{a}_i$, the rules produce one and only one set $\mathbf{a}_i'$. Furthermore, it should be possible to invert the transformation and recover the unique $\mathbf{a}_i$ from the $\mathbf{a}_i'$.

If new basis vectors are chosen, the coordinates of points and the components of vectors must change also. Implicit in our development will be the assumption of a real world consisting of fixed points. We might, for example, describe the location of an atom within a unit cell of a crystal by a unique set of coordinates relative to an initial choice of axes; a change of reference axes does not move the atom, but the numbers that specify its location change. The transformation equations based upon this convention of variable axes in a fixed space express the relationships between two sets of basis vectors and show how the components of a fixed vector depend upon the choice of axes. An alternative interpretation, in which the transformation is regarded as an operator that produces new points or vectors in a space with fixed axes, will be applied to the treatment of symmetry in Chapter 4.

Attention may be focused upon the rules for selecting the new basis vectors, or upon the related rules that express the new components in terms of the old. It is somewhat easier to formulate the general case in terms of component transformations, so we shall start by considering how each x'^i is related to the original components x^1, x^2, x^3. This relationship may be written

$$x'^i = f^i(x^1, x^2, x^3) \tag{3–1}$$

where f^i is the function or rule that shows how x'^i is related to x^1, x^2, x^3. Equation (3–1) indicates that the function generally depends upon the components or coordinates. The condition that the transformation have an inverse, say ϕ^i, where

$$x^i = \phi^i(x'^1, x'^2, x'^3) \tag{3–2}$$

is that the determinant of the transformation

$$J^{-1} = \begin{vmatrix} \dfrac{\partial f^1}{\partial x^1} & \dfrac{\partial f^2}{\partial x^1} & \dfrac{\partial f^3}{\partial x^1} \\ \dfrac{\partial f^1}{\partial x^2} & \dfrac{\partial f^2}{\partial x^2} & \dfrac{\partial f^3}{\partial x^2} \\ \dfrac{\partial f^1}{\partial x^3} & \dfrac{\partial f^2}{\partial x^3} & \dfrac{\partial f^3}{\partial x^3} \end{vmatrix} \tag{3–3}$$

exist and not be zero. The determinant here has been denoted by J^{-1} to suggest that it is the inverse of what is known as the Jacobian of the transformation.

For an infinitesimal change, Eq. (3–1) becomes

$$dx'^i = \frac{\partial f^i}{\partial x^j} dx^j \tag{3–4}$$

which may be written as a matrix equation

$$\mathbf{dx}' = \mathbf{F}\ \mathbf{dx} \tag{3–5}$$

where

$$F^i{}_j = \frac{\partial f^i}{\partial x^j} \tag{3–6}$$

The characteristic feature of transformations in rectilinear spaces is that the elements of **F** are constants, independent of the components of the particular vector, and Eq. (3–5) may be integrated to give

$$\mathbf{x}' = \mathbf{Fx} + \mathbf{t} \tag{3–7}$$

$$x'^i = F^i{}_j\, x^j + t^i \tag{3–8}$$

where the components of vector **t** are integration constants that describe a shift of origin. Equations (3–7) and (3–8) are sufficiently general for our present treatment of rectilinear geometry. The treatment of curvilinear cases in Chapter 7 will require the differential transformation of Eq. (3–5) (hence, the treatment of curvilinear systems is known as differential geometry).

3–2 Shifts of Origin

A pure shift of origin results if **F** in Eq. (3–7) is the unit matrix or identity operator. The equation

$$x'^i = x^i + t^i \tag{3–9}$$

shifts the origin from the point [0, 0, 0] to the point whose original coordinates were $[-t^1, -t^2, -t^3]$. The inverse of this transformation is simply

$$x^i = x'^i - t^i \tag{3-10}$$

Many calculations pertain to vectors between two points, for which

$$x'^i - y'^i = x^i - y^i \tag{3-11}$$

so the shift of origin has no effect. On the other hand, the distance from the new origin to a point will not generally be the same as the distance from the old origin to the same point.

EXERCISE 3-1 In a crystal with $a_1 = 7.00$, $a_2 = 6.00$, $a_3 = 8.00$ Å, $\alpha = 90.0°$, $\beta = 110.0°$, $\gamma = 100.0°$ (see Exercise 2-5), an atom is at 0.150, 0.120, −0.065.

(a) Find the distance of this atom from the origin.

(b) If the origin is shifted to the point whose original coordinates were −1.000, 0, 2.000, calculate the new coordinates of the atom and the distance of the atom from the new origin.

EXERCISE 3-2 Solid chlorine at −160°C is orthorhombic with $a_1 = 6.24$, $a_2 = 8.26$, $a_3 = 4.48$ Å. Two of the eight *Cl* atoms in the unit cell are at 0.250, 0.100, 0.370 and 0.250, 0.900, 0.130. Calculate the length of the *Cl*−*Cl* bond.

EXERCISE 3-3 In the cubic close-packed crystal structure, an atom at the origin of the unit cell is touching an atom at the center of a face. Find the coordinates of all twelve atoms that touch the atom at the origin.

3-3 Transformation of Axes

Since shifts of origin may be accommodated by merely adding or subtracting coordinates, no generality will be lost by ignoring vector **t** in Eq. (3-7) and writing the transformation equation that specifies a new set of axes as

$$\mathbf{x}' = \mathbf{F}\mathbf{x} \tag{3-12}$$

If x^i are the components of a vector relative to an original set of axes $\mathbf{a}_i$, and x'^i are the components of the *same* vector relative to the new axes, we must have

$$\mathbf{x} = x^i\mathbf{a}_i = x'^i\mathbf{a}'_i \tag{3-13}$$

$$\mathbf{x} = \bar{\mathbf{x}}\mathbf{a} = \bar{\mathbf{x}}'\mathbf{a}' \tag{3-14}$$

Equation (3–14) is a matrix equation in which $\bar{\mathbf{x}}\mathbf{a}$ denotes the matrix product of the row matrix with elements x^i with the column matrix whose elements are $\mathbf{a}_i$. A bar over the symbol for a matrix is used generally to indicate the transpose (rows and columns exchanged) of the matrix denoted by the same symbol without the bar. Equations (3–13) and (3–14) are identities showing that we are dealing with two representations of the same vector.

The transpose of Eq. (3–12) is

$$\bar{\mathbf{x}}' = \bar{\mathbf{x}}\bar{\mathbf{F}} \tag{3–15}$$

Hence,

$$\bar{\mathbf{x}}\mathbf{a} = \bar{\mathbf{x}}\bar{\mathbf{F}}\mathbf{a}' \tag{3–16}$$

Let $\mathbf{G}$ be the matrix that transforms the basis vectors; that is,

$$\mathbf{a}' = \mathbf{G}\mathbf{a} \tag{3–17}$$

$$\mathbf{a}_i' = G_i^{\ j}\, \mathbf{a}_j \tag{3–18}$$

Then Eq. (3–16) becomes

$$\bar{\mathbf{x}}\mathbf{a} = \bar{\mathbf{x}}\bar{\mathbf{F}}\mathbf{G}\mathbf{a} \tag{3–19}$$

Equation (3–19) is satisfied if

$$\bar{\mathbf{F}}\mathbf{G} = \mathbf{I} \tag{3–20}$$

where $\mathbf{I}$ is the unit matrix [Eq. (1–38)]. Solving for $\mathbf{G}$,

$$\mathbf{G} = \bar{\mathbf{F}}^{-1} \tag{3–21}$$

or for $\mathbf{F}$

$$\mathbf{F} = \bar{\mathbf{G}}^{-1} \tag{3–22}$$

The condition that transformations be unique ensures that these are the only solutions of Eq. (3–19). Transformations whose matrices are the transpose of the inverse (which is the same as the inverse of the transpose) of each other are said to be *contragredient*. The axis transformation can be determined from the coordinate transformation, and vice versa. Quantities whose transformation properties are given by matrix **G** (with components $G_i{}^j$) are said to be *covariant* (co- means *with*); they transform in the same way as the basis vectors. Those quantities that were labeled with subscripts in Chapter 1 are covariant; in more careful language, they transform covariantly. Quantities transforming in the same way as the coordinates, with transformation matrix **F**, transform contravariantly (contra- means *against*), and these quantities were assigned superscripts in Chapter 1. A way of remembering which type of transformation is associated with which index is to recall that co- rhymes with low, which is the covariant index position.

It is instructive to derive the basis vector transformation also by means of suffix manipulations. Assuming that a transformation matrix **G** exists with $G_k{}^i$ the element in the kth row and ith column,

$$\mathbf{a}_k' = G_k{}^i\,\mathbf{a}_i \tag{3–23}$$

From Eq. (3–8) (with $t^i = 0$) and Eq. (3–13),

$$x'^i\mathbf{a}_i = G_k{}^i\,F^k{}_j\,x^j\mathbf{a}_i \tag{3–24}$$

The coefficient of $\mathbf{a}_i$ must be x'^i on both sides of the equation, so

$$G_k{}^i\,F^k{}_j = \delta^i{}_j \tag{3–25}$$

This is equivalent to

$$\tilde{\mathbf{G}}\mathbf{F} = \mathbf{I} \tag{3–26}$$

The left-to-right order of the indices in such terms as $G_k{}^i$ or $F^j{}_l$ specifies the location of the quantity in the matrix representation; the left-hand index identifies the row, and the right-hand index identifies the column. Thus, $G_1{}^2$ is the second element in the top row of matrix **G** and $F^1{}_2$ is the second element in the top row of matrix **F**. The order of writing the indices is unimportant only if the matrix is symmetric, so that for example

$$A'^{i}{}_{j} = A_{i}{}^{j} \qquad (3\text{-}27)$$

It is desirable also to retain the order of appearance of the indices in certain derived tensors, as, for example,

$$A^{k}{}_{i}{}^{l} = B_{ij}C^{kjl} \qquad (3\text{-}28)$$

(see Sections 3-7 and 3-8 for some special cases of such derived tensors).

EXERCISE 3-4 A transformation of axes is given by

$$\mathbf{G} = \begin{pmatrix} 1 & 1 & 0 \\ 3 & -2 & 0 \\ -1 & -1 & 3 \end{pmatrix}$$

Find **F** and complete the following table:

x^1	x^2	x^3	x'^1	x'^2	x'^3
1	0	0			
0	1	0			
0	0	1			
1	1	1			
			1	0	0
			0	1	0
			0	0	1
			1	1	0

EXERCISE 3-5 Th_2Fe_{17} has been reported to be monoclinic, with $a_1 = 9.68$, $a_2 = 8.56$, $a_3 = 6.46$ Å, $\beta = 99.3°$, space group $C2/m$, Th atoms in positions $4(i)$: $\pm(x, 0, z) + (0, 0, 0; 1/2, 1/2, 0)$, with $x = 0.167$, $z = 0.333$.

(a) Calculate the dimensions of the primitive cell obtained from the monoclinic cell by the transformation

$$\mathbf{G} = \begin{pmatrix} \tfrac{1}{2} & \tfrac{1}{2} & 0 \\ \tfrac{1}{2} & -\tfrac{1}{2} & 0 \\ 0 & 0 & -1 \end{pmatrix}$$

What crystal system is suggested by these dimensions?

(b) Calculate the dimensions of the cell obtained from the monoclinic cell by the transformation

$$\mathbf{G} = \begin{pmatrix} 0 & 1 & 0 \\ \tfrac{1}{2} & -\tfrac{1}{2} & 1 \\ 1 & 0 & -1 \end{pmatrix}$$

What crystal system is suggested by these dimensions? How many lattice points are there in this unit cell?

(c) Calculate the coordinates of all the Th atoms in one of the unit cells defined by the transformation of (a).

(d) Calculate the coordinates of all Th atoms in one of the unit cells defined by the transformation of (b).

EXERCISE 3-6 ScF_3 is rhombohedral, with $a = 4.023$ Å, $\alpha = 89.57°$, (see Exercise 1-23). The Sc atom is at the origin; the three F atoms are at 0.500, 0.027, 0.973; 0.973, 0.500, 0.027; 0.027, 0.973, 0.500.

(a) Transform to a triply-primitive hexagonal cell by

$$\mathbf{G} = \begin{pmatrix} 1 & -1 & 0 \\ 0 & 1 & -1 \\ 1 & 1 & 1 \end{pmatrix}$$

(b) Plot the structure projected onto the hexagonal (001).

3-4 Transformation of Metric Tensor

If a new set of basis vectors is given by Eq. (3-18), the components of the new metric tensor are

$$g'_{kl} = \mathbf{a}'_k \cdot \mathbf{a}'_l \tag{3-29}$$

$$g'_{kl} = (G_k{}^i\, \mathbf{a}_i) \cdot (G_l{}^j\, \mathbf{a}_j) \tag{3-30}$$

$$g'_{kl} = G_k{}^i\, g_{ij} G_l{}^j \tag{3-31}$$

In matrix notation

$$\mathbf{g}' = \mathbf{G}\mathbf{g}\overline{\mathbf{G}} \tag{3-32}$$

An alternative expression of this transformation, in terms of partial derivatives, will be given in Eq. (3-59).

Similarly, the transformation of the reciprocal metric tensor is given by

$$\mathbf{g}'^{*} = \mathbf{F}\mathbf{g}^{*}\overline{\mathbf{F}} \tag{3-33}$$

$$g'^{kl} = F^k{}_i\, g^{ij} F^l{}_j \tag{3-34}$$

EXERCISE 3–7 A crystal has $a_1 = 5.00$, $a_2 = 7.00$, $a_3 = 9.00$ Å. $\alpha = 75.0°$, $\beta = 85.0°$, $\gamma = 120.0°$. A new set of axes is chosen using transformation **G** of Exercise 3–4.

(a) Calculate the dimensions of the new unit cell by using the vector methods of Chapter 1.

(b) Calculate the new metric tensor by taking scalar products of the vectors obtained in (a).

(c) Calculate the new metric tensor by means of Eq. (3–32).

(d) Calculate $\mathbf{g}'^*$ by (i) taking the inverse of the $\mathbf{g}'$ obtained in (c), (ii) calculating $\mathbf{g}^*$ as the inverse of $\mathbf{g}$ and applying Eq. (3–33).

EXERCISE 3–8 Apply transformation **G** of Exercise 3–4 to a tetragonal cell for which $a_1 = a_2 = 6.00$, $a_3 = 8.00$ Å.

(a) Calculate the new metric tensor by applying Eq. (3–32).

(b) Calculate the components of $\mathbf{g}'$ by carrying out the summations implied by Eq. (3–31).

3–5 Unit Cell Volume Relationships

The results of Section 1–9 show that the volume of the unit cell defined by vectors $\mathbf{a}_1$, $\mathbf{a}_2$, $\mathbf{a}_3$ is

$$V = |\mathbf{a}_1 \cdot \mathbf{a}_2 \wedge \mathbf{a}_3| \tag{3–35}$$

Equation (1–25) was presented without proof as a practical formula for calculating the volume. The transformation methods that have been developed now permit simple derivation of Eq. (1–25). Let **G** represent the transformation that generates basis vectors $\mathbf{a}_i$ from a set of cartesian basis vectors $\mathbf{e}_i$. That is,

$$\begin{pmatrix} \mathbf{a}_1 \\ \mathbf{a}_2 \\ \mathbf{a}_3 \end{pmatrix} = \begin{pmatrix} G_1{}^1 & G_1{}^2 & G_1{}^3 \\ G_2{}^1 & G_2{}^2 & G_2{}^3 \\ G_3{}^1 & G_3{}^2 & G_3{}^3 \end{pmatrix} \begin{pmatrix} \mathbf{e}_1 \\ \mathbf{e}_2 \\ \mathbf{e}_3 \end{pmatrix} \tag{3–36}$$

Applying Eq. (3–35) to the vectors produced by Eq. (3–36) gives

$$\begin{aligned} V = {} & G_1{}^1 G_2{}^2 G_3{}^3 - G_1{}^1 G_2{}^3 G_3{}^2 + G_1{}^2 G_2{}^3 G_3{}^1 \\ & - G_1{}^2 G_2{}^1 G_3{}^3 + G_1{}^3 G_2{}^1 G_3{}^2 - G_1{}^3 G_2{}^2 G_3{}^1 \end{aligned} \tag{3–37}$$

But the right-hand side of Eq. (3–37) is just the determinant of **G** (see Exercise 1–52). In transforming from the cartesian system, in which the volume $\mathbf{e}_1 \cdot \mathbf{e}_2 \wedge \mathbf{e}_3$ was unity, to the basis vectors $\mathbf{a}_i$, the volume of the parallelepiped spanned by the basis vectors has been multiplied by the determinant of the transformation matrix. Conversely, in going from $\mathbf{a}_i$ to $\mathbf{e}_i$ by the transformation $\mathbf{G}^{-1}$, the volume is multiplied by the determinant of $\mathbf{G}^{-1}$, which is $1/|\mathbf{G}|$.

In the more general case of transforming from $\mathbf{a}_i$ to $\mathbf{a}_i'$ by

$$\mathbf{a}' = \mathbf{Ma} \tag{3-38}$$

the net result is that of transforming $\mathbf{a}_i$ to a cartesian system by the $\mathbf{G}^{-1}$ used above,

$$\mathbf{e} = \mathbf{G}^{-1}\mathbf{a} \tag{3-39}$$

and then transforming from $\mathbf{e}_i$ to $\mathbf{a}_i'$ by

$$\mathbf{a}' = \mathbf{Te} \tag{3-40}$$

For each of these steps the unit cell volume is multiplied by the determinant of the corresponding matrix, so

$$V' = |\mathbf{T}|\ |\mathbf{G}^{-1}|\, V \tag{3-41}$$

Since

$$\mathbf{M} = \mathbf{TG}^{-1} \tag{3-42}$$

and the determinant of a product of matrices is equal to the product of the determinants, it follows that

$$V' = |\mathbf{M}|\, V = MV \tag{3-43}$$

Thus, for any transformation of axes the unit cell volume is multiplied by the determinant of the transformation matrix.

If Eq. (3-32) is applied to a transformation from a cartesian system [as given by Eq. (3-36)],

$$\mathbf{g}' = \mathbf{GI\bar{G}} \tag{3-44}$$

where **I**, the unit matrix, is the cartesian metric tensor. By Eq. (3–43), V' is equal to $|\mathbf{G}|$ (or G), and we have the important and general result

$$|\mathbf{g}'| = g' = (V')^2 \tag{3–45}$$

By taking the inverse of Eq. (3–44), the reciprocal metric tensor is obtained in terms of the matrix that transforms coordinates from a cartesian to the general system:

$$\mathbf{g}'^* = \mathbf{F}\mathbf{I}\bar{\mathbf{F}} \tag{3–46}$$

And, by analogy with Eq. (3–45),

$$|\mathbf{g}'^*| = g'^* = (V'^*)^2 \tag{3–47}$$

Two very valuable results have been derived in this section. The first is that a transformation multiplies volume by the determinant of the transformation matrix. This result applies even to curvilinear systems in which the differential volume element is multiplied by the Jacobian J of the transformation; J corresponds to the determinant of **G**. This correspondence will be explored in greater depth in Chapter 7 [see also Eq. (3–3)]. Our second important result here is that the determinant of the metric tensor is equal to the square of the unit cell volume.

EXERCISE 3–9 Derive Eq. (1–25) by evaluating the determinant of the metric tensor.

EXERCISE 3–10 Given the basis vector transformations,

$$\mathbf{a}_i' = T_i^{\ j}\,\mathbf{a}_j, \qquad \mathbf{a}_i'' = U_i^{\ j}\,\mathbf{a}_j'$$

where

$$\mathbf{T} = \begin{pmatrix} 4 & 2 & 0 \\ 1 & 1 & -1 \\ 0 & 1 & -1 \end{pmatrix} \qquad \mathbf{U} = \begin{pmatrix} 1 & -1 & 1 \\ 0 & 3 & -2 \\ 1 & -2 & 2 \end{pmatrix}$$

fill in the following table

x^1	x^2	x^3	x'^1	x'^2	x'^3	x''^1	x''^2	x''^3
0.50,	0.50,	0.00						
0.00,	0.00,	0.00						
			1.00,	−1.00,	0.00			
						3.00,	0.50,	0.50

EXERCISE 3-11 If the original set of axes in Exercise 3-10 was cartesian, calculate $\mathbf{g}'$, $\mathbf{g}''$, V', and V''.

EXERCISE 3-12 Calculate the inverse of $\mathbf{g}''$ of Exercise 3-11 and hence determine the reciprocal cell dimensions.

EXERCISE 3-13 Given a crystal with metric tensor $\mathbf{g}$, derive the transformations $\mathbf{G}$ and $\mathbf{F}$ to a set of cartesian axes $\mathbf{e}_i$ in which $\mathbf{e}_1$ is parallel to $\mathbf{a}_1$, and $\mathbf{e}_2$ is in the $\mathbf{a}_1$-$\mathbf{a}_2$ plane. (Hint: Note that $\mathbf{e}_3$ is parallel to $\mathbf{a}^3$.)

EXERCISE 3-14 Use the result of Exercise 3-13 to derive the matrices $\mathbf{G}$ and $\mathbf{F}$ for the transformation to a cartesian coordinate system of a crystal for which $a_1 = 5.50$, $a_2 = 6.20$, $a_3 = 7.14$ Å, $\alpha = 99.0°$, $\beta = 107.5°$, $\gamma = 101.4°$.

EXERCISE 3-15 Suppose $\mathbf{G}$ represents the transformation from one set of cartesian axes to another. Show that $G = \pm 1$ and that $\mathbf{G} = \bar{\mathbf{G}}^{-1}$. (A transformation with these properties is said to be orthogonal.)

EXERCISE 3-16 Show that the sum of the diagonal elements of a second-rank tensor (the trace) is invariant under an orthogonal transformation (see Exercise 3-15).

EXERCISE 3-17 Suppose that tensors **A**, **B**, and **C** have components A^{ij}, $B^i{}_j$, and C_{ij} in a cartesian coordinate system. Transform to a general rectilinear coordinate system by transformation matrices **F** and **G**, and deduce the invariant relationships in the general system that correspond to the trace in a cartesian system.

EXERCISE 3-18 If **A** is a second-rank tensor with components A^{ij}, show that $A^{11}A^{22} + A^{22}A^{33} + A^{33}A^{11} - A^{12}A^{21} - A^{23}A^{32} - A^{31}A^{13}$ is invariant under an orthogonal transformation. This quantity is called the second invariant, whereas the trace itself is called the first invariant. An application of this to thermal expansion tensors has been given by D. Weigel, T. Beguemsi, P. Garnier, and J. F. Berar, J. Solid State Chem., **23**, 241-251 (1978).

EXERCISE 3-19 Consider tensors **A**, **B**, and **C**, with components A^{ij}, $B^i{}_j$, and C_{ij}.

(a) If these tensors are symmetric in some coordinate system, show that **A** and **C**, but not **B**, are symmetric in any coordinate system.

(b) If these tensors are antisymmetric in some coordinate system, show that **A** and **C**, but not **B**, are antisymmetric in any coordinate system.

3-6 Tensor Definitions

We have seen that the coordinates of points, which we label with superscripts and refer to as contravariant, transform by Eq. (3-4), which in rectilinear spaces with no origin shift integrates to

$$x'^{i} = F^{i}_{\ j}\, x^{j} \tag{3–48}$$

The quantities $F^{i}_{\ j}$ are the partial derivatives of the new coordinates with respect to the old:

$$F^{i}_{\ j} = \frac{\partial x'^{i}}{\partial x^{j}} \tag{3–49}$$

The basis vectors, which we label with subscripts and refer to as covariant, transform by Eq. (3–18), in which the elements $G_i^{\ j}$ define a matrix **G** that is contragedient to **F**; that is, **G** is equal to the transpose of the inverse of **F**. Since the inverse of **F** will give the original coordinates x^i in terms of the x'^{i}, it follows that

$$G_i^{\ j} = \frac{\partial x^{j}}{\partial x'^{i}} \tag{3–50}$$

A *tensor* is a set of quantities that may be transformed by a product of transformations of the type given by Eqs. (3–49) and (3–50). Quantities that do not transform according to these equations are not tensors.

The rank (or order) of a tensor is equal to the number of transformations of the type of Eqs. (3–49) and (3–50) that must be applied to obtain the new values. A tensor of rank r in a space of dimension D has D^r components.

A *scalar* may be regarded as a tensor of rank zero. Its value is invariant under transformation.

The components of vectors are first-rank contravariant tensors, and the basis vectors are first-rank covariant tensors. The reciprocal basis vectors transform in the same way as coordinates and are therefore first-rank contravariant tensors.

If a vector **v** is referred to reciprocal basis vectors $\mathbf{a}^i$,

$$\mathbf{v} = h_i \mathbf{a}^i \tag{3–51}$$

where the h_i are written with subscripts in anticipation of their covariance. In order to demonstrate that the h_i are indeed covariant, consider the expression of vector **v** referred to two different sets of reciprocal basis vectors:

$$\mathbf{v} = h_i' \mathbf{a}'^{i} = h_j \mathbf{a}^j \tag{3–52}$$

The transformation equation for the contravariant $\mathbf{a}^i$ is

$$\mathbf{a}'^{i} = F^{i}{}_{j}\,\mathbf{a}^{j} \tag{3-53}$$

Hence,

$$h_i' F^{i}{}_{j}\,\mathbf{a}^{j} = h_j \mathbf{a}^{j} \tag{3-54}$$

It is apparent that this equation is satisfied if

$$h_i' F^{i}{}_{j} = h_j \tag{3-55}$$

and the requirement that transformations be unique ensures that this is the only solution of Eq. (3-54). In matrix notation

$$\bar{\mathbf{h}}'\mathbf{F} = \bar{\mathbf{h}} \tag{3-56}$$

where the bar denotes a row vector, as opposed to the column vector $\mathbf{h}$. Multiplying both sides of Eq. (3-56) on the right by $\mathbf{F}^{-1}$ yields

$$\bar{\mathbf{h}}' = \bar{\mathbf{h}}\mathbf{F}^{-1} \tag{3-57}$$

and taking the transpose produces

$$\mathbf{h}' = \bar{\mathbf{F}}^{-1}\mathbf{h} = \mathbf{G}\mathbf{h} \tag{3-58}$$

Thus, the invariance of $h_j \mathbf{a}^j$ under transformation leads to the result that the components of vectors in reciprocal space are first-rank covariant tensors.

Equation (2-39) described a plane by means of

$$c x^i h_i = 1$$

Reasoning similar to that used above shows that the h_i, the Miller indices of the plane, transform as first-rank covariant tensors.

The metric tensor transformations of Eqs. (3-32) and (3-33) clearly satisfy the tensor definition and justify the names that we have applied to **g** and **g*** since Chapter 1. The metric tensor **g** is a second-rank covariant tensor, and an alternative expression of Eq. (3-31) is

$$g'_{kl} = \frac{\partial x^i}{\partial x'^k}\,\frac{\partial x^j}{\partial x'^l}\,g_{ij} \tag{3-59}$$

Likewise, the reciprocal metric tensor is second-rank contravariant, and an alternative expression of Eq. (3-34) is

$$g'^{kl} = \frac{\partial x'^k}{\partial x^i} \frac{\partial x'^l}{\partial x^j} g^{ij} \qquad (3\text{–}60)$$

Important particular cases of Eqs. (3–59) and (3–60) result if the original system is cartesian.

These results may be generalized to tensors of any rank, even those with mixtures of subscripts and superscripts. For example, the quantity **A** with transformation properties

$$A'^i{}_{jk} = F^i{}_l \, G_j{}^m \, G_k{}^n \, A^l{}_{mn} \qquad (3\text{–}61)$$

or

$$A'^i{}_{jk} = \frac{\partial x'^i}{\partial x^l} \frac{\partial x^m}{\partial x'^j} \frac{\partial x^n}{\partial x'^k} A^l{}_{mn} \qquad (3\text{–}62)$$

is a third-rank tensor with one contravariant and two covariant indices. Elementary matrix operations are not convenient for tensors of rank higher than two, but the index notation and the summation convention still apply, and the right-hand side of Eq. (3–61) or Eq. (3–62) represents 27 terms.

EXERCISE 3–20 Fill in the following table for the Miller indices of planes in a crystal of Th_2Fe_{17}, where the h_i are the indices for the C-centered monoclinic cell, the h_i' are for the primitive cell given by the transformation of Exercise 3–5(a), and the h_i'' are the indices for the triply primitive cell given by the transformation of Exercise 3–5(b). What conditions on reflections pertain to the three cases?

h_1	h_2	h_3	h_1'	h_2'	h_3'	h_1''	h_2''	h_3''
1	3	2						
2	4	0						
			1	2	4			
						$\bar{1}$	1	1
						0	0	3

3–7 Raising and Lowering Indices

The ability of the metric tensors to raise and lower indices was noted in Chapter 1. Thus, g_{ij} generates $\mathbf{a}_i$ from $\mathbf{a}^j$ [Eq. (1–41)] or v_i from v^j [Eq. (1–52)], and g^{ij} *generates* $\mathbf{a}^i$ from $\mathbf{a}_j$ [Eq. (1–35)] or v^i from v_j [Eq. 1–53]. It is generally true that g_{ij} will lower an index and g^{ij} will raise an index. Consider, for example, a tensor **A** with components $A_y{}^{jx}$, where j is a contravariant index and x and y represent whatever strings of contravariant and covariant indices are necessary to complete the description. By Eq. (3–31)

$$g'_{kj}A'^{jx}_{y} = G_k{}^m\, g_{mn} G_j{}^n\, A'^{jx}_{y} \tag{3-63}$$

Since j is a contravariant index of **A**,

$$A'^{jx}_{y} = F^j{}_l\, A_y{}^{lx} \tag{3-64}$$

The transformation matrices affecting the x and y indices are irrelevant to the present argument and so have been omitted from Eq. (3–64). Putting Eq. (3–64) into the right-hand side of Eq. (3–63)

$$g'_{kj}A'^{jx}_{y} = G_k{}^m\, g_{mn} G_j{}^n\, F^j{}_l\, A_y{}^{lx} \tag{3-65}$$

Using Eq. (3–25) produces

$$g'_{kj}A'^{jx}_{y} = G_k{}^m\, g_{mn}\delta^n{}_l\, A_y{}^{lx} \tag{3-66}$$

which reduces to

$$g'_{kj}A'^{jx}_{y} = G_k{}^m\, g_{mn}A_y{}^{nx} \tag{3-67}$$

Thus, **G** is the appropriate transformation for **gA**, and, since **G** transforms covariant quantities, the product **gA** has one fewer contravariant and one more covariant index than **A**, and the effect of **g** has been to lower an index. We may write

$$A_{yi}{}^x = g_{ij}A_y{}^{jx} \tag{3-68}$$

Quantities related by raised or lowered suffixes, such as $A_{yi}{}^x$ and $A_y{}^{jx}$, are known as *associated* tensors.

EXERCISE 3–21 Suppose a tensor **B** has a covariant index. Carry out a calculation parallel to that of Eqs. (3–63) to (3–68) to show that **g*B** is contravariant.

EXERCISE 3–22 Apply the raising operator **g*** to the tensor **g** to obtain the mixed tensor with components $g^i{}_j$.

3–8 Contraction of Indices

Suppose that C^{ij} are the components of a second-rank contravariant tensor and D_{kl} are the components of a second-rank covariant tensor. The product $C'^{ij}\, D'_{jk}$ is given by

$$C'^{ij}D'_{jk} = F^i{}_l\, F^j{}_m\, C^{lm} G_j{}^n\, G_k{}^p\, D_{np} \qquad (3\text{–}69)$$

$$C'^{ij}D'_{jk} = F^i{}_l\, G_k{}^p\, \delta_m{}^n\, C^{lm} D_{np} \qquad (3\text{–}70)$$

$$C'^{ij}D'_{jk} = F^i{}_l\, G_k{}^p\, C^{lm} D_{mp} \qquad (3\text{–}71)$$

This shows that $C^{ij}D_{jk}$ is a component of a tensor that has one covariant and one contravariant index. Denoting this tensor by **B**,

$$B^i{}_k = C^{ij}D_{jk} \qquad (3\text{–}72)$$

Thus, summation over an index that appears in both the superscript and the subscript position produces a tensor with the common index eliminated; the operation has produced *contraction* of the tensor. The raising and lowering of indices by means of the metric tensors are in fact special cases of contraction.

Exercise 3–23 The permutation tensors were defined in Section 1–14. Starting with ϵ_{ijk}, show that

$$\epsilon^{lmn} = g^{li} g^{mj} g^{nk} \epsilon_{ijk}$$

and verify by actual summation the values of ϵ^{123}, ϵ^{132}, and ϵ^{122}.

Exercise 3–24 Calculate all the components of $\epsilon^1{}_{ij}$, $\epsilon^2{}_{ij}$, and $\epsilon^3{}_{ij}$.

3–9 Some Properties of Tensors

The transformation properties that have been developed lead to the following properties.

If **B** is a tensor and b is a scalar, $b\mathbf{B}$ is a tensor.

If **A** and **B** are tensors of the same covariant and contravariant rank, the sum $\mathbf{A} + \mathbf{B}$ is a tensor of the same rank.

The product of two tensors with different indices gives a new tensor whose rank is the sum of the two individual ranks. Thus,

$$A^{ij}{}_k\, B^l{}_m = C^{ij}{}_{km}{}^l \qquad (3\text{–}73)$$

If, however, a subscript and a superscript are the same, contraction reduces the rank of the tensor by two. As a special case of Eq. (3–73), the products of coordinates form tensors; $x^i x^j$ is a second-rank contravariant tensor.

Equation (3–73) may be regarded as an *operator* equation,

$$\mathbf{AB} = \mathbf{C} \tag{3-74}$$

in which tensor **A** operates on tensor **B** to produce the new tensor **C**. Equation (3–68) is of this form, with the metric tensor operating on a tensor to produce its associated tensor. Another special case of Eq. (3–74) is

$$\mathbf{Au} = \mathbf{v} \tag{3-75}$$

in which tensor **A** operates on vector **u** to produce vector **v**.

The structure of tensor equations should be noted carefully. An equation that represents a physical law must have consistent tensor character. While the form $u^i u_i$ produces a scalar $(u)^2$ equal to the square of the length of vector u, the form $u^i u^i$ is the sum of the squares of the contravariant components and is not invariant under transformation. Given the u^i, the combination $u^i g_{ij} u^j$ contracts to produce $(u)^2$. Similarly, the form $h_i g^{ij} h_j$ has the scalar tensor character demanded of $(h)^2$. It is advisable to be suspicious of any equations that do not display tensor consistency.

The formulation of a physical law as a tensor equation is independent of the coordinate system. If, for example, the law is

$$\mathbf{A} = \mathbf{0} \tag{3-76}$$

where **A** is a tensor of some complexity and **0** is the null tensor (with all components zero), transformation to another coordinate system produces

$$\mathbf{A}' = \mathbf{0} \tag{3-77}$$

One of the principal uses of tensor analysis is to provide generalized expressions of physical laws that are valid in any coordinate systems.

EXERCISE 3–25 The coordinates of a point are $x^1 = 0.3200$, $x^2 = 0.1200$, $x^3 = -0.4000$.

(a) Calculate all the products $x^i x^j$.

(b) A new set of axes is chosen by the transformation

$$\mathbf{G} = \begin{pmatrix} 1 & 2 & 0 \\ 0 & 1 & -1 \\ 3 & -1 & 1 \end{pmatrix}$$

Calculate matrix **F**, and determine the new coordinates x'^1, x'^2, x'^3.

(c) From the results of (b) calculate all products $x'^i x'^j$.

(d) Calculate the products $x'^i x'^j$ by transforming the products $x^i x^j$ determined in (a) as a second-order contravariant tensor.

EXERCISE 3–26 The term *covariance* has a different meaning in statistics, where it denotes the average value of the product of the deviations of two quantities from their respective means. That is,

$$\mathrm{cov}\,(s,\,t) \;=\; <(s - \bar{s})(t - \bar{t})>$$

where both the bars and the brackets indicate mean values. If $s = t$, the result is called the variance:

$$\mathrm{var}\,(s) \;=\; \mathrm{cov}\,(s,s) \;=\; \sigma^2(s) \;=\; <(s - \bar{s})^2>$$

Let **V** be the variance–covariance matrix of the coordinates of atoms, with elements $V^{ij} = \mathrm{cov}(x^i, x^j)$ (**V** is generated in the process of refining atomic parameters by the method of least squares). Show that **V** is a second-rank contravariant tensor (see Sands, 1966).

3–10 Derivatives of Tensors

Let $f(x^1, x^2, x^3)$ be a scalar function of the coordinates x^i. The chain rule for partial differentiation is

$$\frac{\partial f}{\partial x'^i} = \frac{\partial x^j}{\partial x'^i}\,\frac{\partial f}{\partial x^j} \tag{3–78}$$

$$\frac{\partial f}{\partial x'^i} = G_i^{\;j}\,\frac{\partial f}{\partial x^j} \tag{3–79}$$

Thus, the derivatives with respect to x^j of a scalar function transform as covariant vector components; similarly, the derivatives of a scalar function with respect to covariant components produce contravariant vector components. Representing the covariant components in Eq. (3–79) by w_j,

$$w_i' = G_i^{\;j}\, w_j \tag{3–80}$$

Differentiation of Eq. (3–80) with respect to x'^k gives

$$\frac{\partial w_i'}{\partial x'^k} = \frac{\partial G_i^{\,j}}{\partial x'^k} w_j + G_i^{\,j} \frac{\partial w_j}{\partial x'^k} \tag{3–81}$$

$$\frac{\partial w_i'}{\partial x'^k} = \frac{\partial G_i^{\,j}}{\partial x^l} \frac{\partial x^l}{\partial x'^k} w_j + G_i^{\,j} \frac{\partial w_j}{\partial x^l} \frac{\partial x^l}{\partial x'^k} \tag{3–82}$$

Therefore,

$$\frac{\partial w_i'}{\partial x'^k} = G_k^{\,l} \frac{\partial G_i^{\,j}}{\partial x^l} w_j + G_i^{\,j} G_k^{\,l} \frac{\partial w_j}{\partial x^l} \tag{3–83}$$

Equation (3–83) reveals that the derivatives of covariant components with respect to contravariant components generally do not form second-rank covariant tensors. In rectilinear systems, however, in which the elements of the transformation matrices are independent of the coordinates, the first term on the right of Eq. (3–83) vanishes, and derivatives do obey tensor transformation rules. We shall return to the subject of derivatives in Chapter 7.

3–11 Transformation of Permutation Tensors

According to Exercise 1–52, the determinant of the 3×3 matrix $\mathbf{G}$ is

$$|\mathbf{G}| = J = G_1^{\,i} G_2^{\,j} G_3^{\,k} e_{ijk} \tag{3–84}$$

where the symbol J is used as a reminder that this particular determinant is the Jacobian of the transformation [see Eq. (3–3)]. Multiplying both sides by the permutation symbol e_{lmn} gives

$$Je_{lmn} = G_1^{\,i} G_2^{\,j} G_3^{\,k} e_{ijk} e_{lmn} \tag{3–85}$$

By virtue of the property of determinants that exchange of two rows changes its sign, Eq. (3–85) is equivalent to

$$e_{lmn} = J^{-1} G_l^{\,i} G_m^{\,j} G_n^{\,k} e_{ijk} \tag{3–86}$$

Permutation symbols are independent of the coordinate system; thus,

$$e'_{lmn} = J^{-1} G_l{}^i \, G_m{}^j \, G_n{}^k \, e_{ijk} \tag{3–87}$$

The factor J^{-1} in Eq. (3–87) prevents e_{ijk} from transforming like a tensor. Quantities whose transformations differ from those of tensors by a factor J^w are sometimes called *relative tensors of weight* w (see, for example, Aris, 1962). The true tensors, for which $w = 0$, are then called *absolute tensors.* Thus, e_{ijk} is a third-rank relative covariant tensor of weight -1. Similarly, e^{ijk} is a third-rank relative contravariant tensor of weight $+1$.

Multiplying both sides of Eq. (3–87) by $\mathbf{a}'_1 \cdot \mathbf{a}'_2 \wedge \mathbf{a}'_3 = P'V'$ [see Eq. (1–57)] produces the permutation tensor

$$\epsilon'_{lmn} = P'V'e'_{lmn} = J^{-1} G_l{}^i \, G_m{}^j \, G_n{}^k \, e_{ijk} P'V' \tag{3–88}$$

But

$$P'V' = JPV \tag{3–89}$$

Hence,

$$\epsilon'_{lmn} = G_l{}^i \, G_m{}^j \, G_n{}^k \, \epsilon_{ijk} \tag{3–90}$$

This proves that ϵ_{ijk} is an absolute third-rank covariant tensor. From the determinant property

$$\epsilon'_{lmn} = J\epsilon_{lmn} \tag{3–91}$$

Similarly, ϵ^{lmn} is a third-rank contravariant tensor, and

$$\epsilon'^{lmn} = J^{-1}\epsilon^{lmn} \tag{3–92}$$

The sign reversal resulting from negative Jacobians should be noted.

3–12 Transformation of Vector Product

By Eq. (1–83),

$$\mathbf{u} \wedge \mathbf{v} = \epsilon_{ijk} u^i v^j g^{kl} \mathbf{a}_l \tag{3–93}$$

where the terms

$$w^l = \epsilon_{ijk} u^i v^j g^{kl} \tag{3–94}$$

are ostensibly the contravariant components of the vector **u** Λ **v**. To verify that the w^l are indeed contravariant, we transform each of the quantities on the right-hand side of Eq. (3–94):

$$w'^l = J\epsilon_{ijk} F^i{}_n F^j{}_o F^k{}_p F^l{}_q u^n v^o g^{pq} \tag{3–95}$$

$$w'^l = J\epsilon_{nop} J^{-1} F^l{}_q u^n v^o g^{pq} \tag{3–96}$$

$$w'^l = F^l{}_q \epsilon_{nop} u^n v^o g^{pq} \tag{3–97}$$

$$w'^l = F^l{}_q w^q \tag{3–98}$$

The steps from Eq. (3–95) to Eq. (3–98) have made use of Eq. (3–91), of the transformation properties of coordinates and of **g***, and of the fact that the determinant of **F** is J^{-1}.

Thus, w^l is a contravariant vector component, and the vector product **u** Λ **v** apparently transforms like any other vector. Nevertheless, a subtle distinction should be noted, and many authors refer to **u** Λ **v** as an *axial vector* or a *pseudovector* to distinguish it from the ordinary *polar* vectors. The difference arises when the transformations have negative Jacobians. The formulation used here includes the negative sign in the permutation tensor as the factor P (parity) that was introduced in Section 1–14. Inclusion of P means, for example, that

$$\mathbf{a}^3 = P(\mathbf{a}_1 \Lambda \mathbf{a}_2)/V \tag{3–99}$$

which assures that left-handed axes produce left-handed reciprocal axes, as mandated by the definition of Eq. (1–26). Without the factor P, ϵ_{ijk} does not transform as a tensor. In cartesian spaces it is tempting to identify the permutation tensor with the permutation symbol, but the change of sign on change of handedness cannot be overlooked. Thus, whereas $\mathbf{e}_1 \Lambda \mathbf{e}_2 = \mathbf{e}_3$ for a right-handed cartesian system, $\mathbf{e}_1 \Lambda \mathbf{e}_2 = -\mathbf{e}_3$ if the system is left-handed. The crucial point here is that transformations with negative Jacobians change the handedness of the basis vectors; the transformation equations give the correct results as long as the new components are properly referred to basis vectors of opposite handedness, but a sign correction is necessary if both basis vector sets are assumed to have the same handedness.

A more substantial difference between vector products and polar vectors will be explored in Section 4–11, which describes the effect on $\mathbf{u} \wedge \mathbf{v}$ of symmetry operations, which leave the basis vectors and the permutation tensor unchanged.

EXERCISE 3–27 A right-handed cartesian coordinate system has

$\mathbf{u} = [5, 3, -2]$, $\mathbf{v} = [6, 1, 4)$.

(a) Calculate $\mathbf{u} \wedge \mathbf{v}$.

(b) A transformation

$$\mathbf{G} = \begin{pmatrix} 0 & 1 & 0 \\ \bar{1} & 0 & 0 \\ 0 & 0 & \bar{1} \end{pmatrix}$$

is applied. (i) Calculate $\mathbf{u}'$, $\mathbf{v}'$, and $\mathbf{u}' \wedge \mathbf{v}'$. (ii) Calculate $\mathbf{u}' \wedge \mathbf{v}'$ by transforming the result of (a).

EXERCISE 3–28 A crystal has $g_{11} = 140$, $g_{12} = -20$, $g_{13} = 30$, $g_{22} = 170$, $g_{23} = 50$, $g_{33} = 200$. Vectors $\mathbf{u}$ and $\mathbf{v}$ are given by $\mathbf{u} = [0.300, 0.700, -0.500]$, $\mathbf{v} = [-0.200, 0.900, 1.100]$. The coordinate system is right-handed.

(a) Calculate the contravariant components of $\mathbf{u} \wedge \mathbf{v}$.

(b) A transformation

$$\mathbf{G} = \begin{pmatrix} 2 & 1 & 0 \\ 0 & -1 & 1 \\ -1 & 0 & 1 \end{pmatrix}$$

is applied. (i) Calculate $\mathbf{u}'$ and $\mathbf{v}'$ and then $\mathbf{u}' \wedge \mathbf{v}'$. (ii) Calculate $\mathbf{u}' \wedge \mathbf{v}'$ by transforming $\mathbf{u} \wedge \mathbf{v}$ found in (a).

3–13 Anisotropic Temperature Factors

For a crystal in which the atoms are at rest, the structure factor is given by

$$F(\mathbf{h}) = \sum_J f_J \exp[2\pi i(\mathbf{h} \cdot \mathbf{r}_J)] \qquad (3\text{–}100)$$

in which f_J is the form factor or atomic scattering factor of atom J, whose coordinates are r_J^1, r_J^2, r_J^3. The J summation is over all atoms contained in one unit cell. Thermal agitation will cause the atoms to fluctuate about their mean positions, so $\mathbf{r}_J$ should be replaced in Eq. (3–100) by its instantaneous value $\mathbf{r}_J + \mathbf{u}_J$. When this is inserted into the Eq. (3–100) series, expansion of the exponential factor $\exp[2\pi i(\mathbf{h} \cdot \mathbf{u})]$ followed by averaging of products produces

$$F(\mathbf{h}) = \sum_J f_J \exp[2\pi i(\mathbf{h} \cdot \mathbf{r}_J)] \exp[-\beta_J{}^{kl} h_k h_l] \qquad (3\text{–}101)$$

where

$$\beta_J{}^{kl} = 2\pi^2 \overline{u_J{}^k u_J{}^l} \qquad \text{(no summation)} \qquad (3\text{–}102)$$

The derivation of Eq. (3–101) neglected terms higher than the quadratic in products of displacements. For simplicity, the atom designation J will be omitted in the subsequent discussion.

The condition that $\beta^{kl}h_k h_l$ be invariant under transformation implies that the β^{kl} are contravariant tensor components with transformation properties

$$\beta'^{kl} = F^k{}_i F^l{}_j \beta^{ij} \qquad (3\text{–}103)$$

$$\beta' = \mathbf{F}\beta\overline{\mathbf{F}} \qquad (3\text{–}104)$$

The averages $\overline{u^k u^l}$ of course also form a tensor

$$U^{kl} = \overline{u^k u^l} = \beta^{kl}/2\pi^2 \qquad (3\text{–}105)$$

The invariance condition

$$\beta^{ij}h_i h_j = \overline{\mathbf{h}}\beta\mathbf{h} = \text{constant} \qquad (3\text{–}106)$$

defines a quadric surface, which represents the second-rank tensor. If β is positive-definite [$|\beta|$, β^{II}, $\beta^{II}\beta^{JJ} - \beta^{IJ}\beta^{IJ}$ (no summation), all positive], the representation surface is an ellipsoid. If the motions are isotropic, the representation surface is a sphere and in cartesian coordinates β is proportional to $\mathbf{I}$. In particular,

$$\beta(\text{isotropic, cartesian}) = 2\pi^2\overline{(u)^2}\mathbf{I} \qquad (3\text{–}107)$$

In the general coordinate system, therefore, Eq. (3–104) and (3–33) show that **β** for an atom undergoing isotropic motion is proportional to **g***. From Eq. (3–107), we may write

$$\beta^{kl}(\text{isotropic}) = 2\pi^2\overline{(\mathrm{u})^2}g^{kl} \quad (3\text{–}108)$$

and the temperature correction to the structure factor is

$$\begin{aligned} T &= \exp[-\beta^{kl}(\text{isotropic})\, h_k h_l] \\ &= \exp[-2\pi^2\, \overline{(u)^2} g^{kl} h_k h_l] \end{aligned} \quad (3\text{–}109)$$

Equation (2–40) identified $g^{kl}h_kh_l$ with $1/(d)^2$, where d is the interplanar spacing. By Bragg's law, d is related to the diffraction angle θ by

$$\lambda = 2\, d \sin\theta \quad (3\text{–}110)$$

Hence,

$$T = \exp[-B(\sin\theta\, /\, \lambda)^2] \quad (3\text{–}111)$$

where B, the isotropic temperature factor, is defined by

$$B = 8\pi^2\, \overline{(u)^2} \quad (3\text{–}112)$$

From Eq. (3–108),

$$\beta^{kl}(\text{isotropic}) = Bg^{kl}/4 \quad (3\text{–}113)$$

EXERCISE 3–29 A certain atom in a crystal has $\beta^{11} = 0.2400$, $\beta^{12} = 0.0200$, $\beta^{13} = -0.0100$, $\beta^{22} = 0.3600$, $\beta^{23} = 0.1000$, $\beta^{33} = 0.3200$. A new set of axes is chosen by the transformation of Exercise 3–4. Calculate the new values of β^{ij}.

EXERCISE 3–30 An atom in a crystal has $\beta^{11} = 0.00515$, $\beta^{12} = 0.00101$, $\beta^{13} = 0.00125$, $\beta^{22} = 0.00214$, $\beta^{23} = -0.00013$, $\beta^{33} = 0.01383$. The crystal is triclinic with $a_1 = 13.01$, $a_2 = 18.83$, $a_3 = 8.32$ Å, $\alpha = 100.40°$, $\beta = 102.12°$, $\gamma = 105.19°$. A cartesian system is chosen by the transformation of Exercise 3–13. Calculate the components of **β** in this cartesian system.

EXERCISE 3–31 Calculate the β^{ij} corresponding to the isotropic $B = 2.50$ Å^2 in a monoclinic crystal with $a_1 = 6.00$, $a_2 = 8.00$, $a_3 = 5.00$ Å, $\beta = 109.0°$.

EXERCISE 3-32 Some crystallographers use a temperature factor with components W_{ij} that are related to the β^{ij} by $W_{IJ} = \beta^{IJ}/(2\pi^2 a^I a^J)$ (no summation). No special tensor character is meant to be implied by the positions assigned here to the indices on **W**. [These authors use the symbol **U** rather than **W**, but we have reserved **U** for the tensor described by Eq. (3–105).] (a) How does **W** transform? (b) Is **W** a tensor?

3–14 Principal Axes of Second-Rank Tensors

As was pointed out in Section 3–13, a second-rank tensor may be represented by a quadric surface. If the tensor is **T**, points on this surface are vectors **v** for which the quadratic form $\bar{\mathbf{v}}\,\mathbf{T}\,\mathbf{v}$ is constant. If the matrix **T** is positive-definite, the representation surface is an ellipsoid. The principal axes of an ellipsoid are extrema vectors, which are invariant in direction upon operation by the tensor (see Waser, 1955; Busing and Levy, 1958; Sands, 1966). That is, the principal axes of tensor **T** are vectors **v** such that

$$\mathbf{T}\mathbf{v} = \lambda\mathbf{v} \tag{3–114}$$

where λ is a constant known as the principal value or eigenvalue of the tensor. If **T** is doubly contravariant,

$$T^{ij}v_j = \lambda v^i \tag{3–115}$$

Solving Eq. (3–115) for the components of the vector **v** requires that the same tensor character for these components be used on both sides. Hence,

$$T^{ij}g_{jk}v^k = \lambda v^i \tag{3–116}$$

or

$$T^{ij}v_j = \lambda g^{ik}v_k \tag{3–117}$$

Nontrivial solutions for the principal vectors **v** (also known as the characteristic vectors or eigenvectors) and for the eigenvalues λ may be obtained from the secular equation

$$|T^{ij}g_{jk} - \lambda\delta^i{}_k| = 0 \tag{3–118}$$

or from the equivalent formulation

$$|T^{ij} - \lambda g^{ij}| = 0 \tag{3–119}$$

In matrix notation these equations are

$$|\mathbf{Tg} - \lambda\mathbf{I}| = 0 \tag{3–120}$$

$$|\mathbf{T} - \lambda\mathbf{g}^*| = 0 \tag{3–121}$$

The following example will illustrate the use of these equations. Suppose that an atom in a monoclinic crystal has anisotropic temperature factor components $\beta^{11} = 0.00400$, $\beta^{12} = 0.00100$, $\beta^{13} = -0.00050$, $\beta^{22} = 0.00300$, $\beta^{23} = 0.00070$, $\beta^{33} = 0.00500$. The unit cell dimensions are $a_1 = 8.00$, $a_2 = 10.00$, $a_3 = 9.00$ Å, $\beta = 105.0°$. We wish to determine the principal axes of the ellipsoid that represents this tensor. Equation (3–120) gives the secular equation

$$\left| \begin{pmatrix} 0.00400 & 0.00100 & -0.00050 \\ 0.00100 & 0.00300 & 0.00070 \\ -0.00050 & 0.00070 & 0.00500 \end{pmatrix} \begin{pmatrix} 64.00 & 0 & -18.63 \\ 0 & 100.0 & 0 \\ -18.63 & 0 & 81.00 \end{pmatrix} - \lambda \begin{pmatrix} 1 & 0 & 0 \\ 0 & 1 & 0 \\ 0 & 0 & 1 \end{pmatrix} \right| = 0$$

Expansion of the determinant produces the cubic equation

$$0.0250 - 0.2917\lambda + 0.9796\lambda^2 - \lambda^3 = 0$$

which has solutions $\lambda = 0.4817, 0.3500, 0.1481$. Equation (3–116) may now be used to find the eigenvectors corresponding to each of the eigenvalues. The first root, for example, leads to the matrix equation

$$\begin{pmatrix} 0.2653 & 0.1000 & -0.1150 \\ 0.0510 & 0.3000 & 0.0381 \\ -0.1252 & 0.0700 & 0.4143 \end{pmatrix} \begin{pmatrix} v^1 \\ v^2 \\ v^3 \end{pmatrix} = 0.4817 \begin{pmatrix} v^1 \\ v^2 \\ v^3 \end{pmatrix}$$

The three simultaneous equations for v^i are not linearly independent. However, two components can be found in terms of the third, and normalization to unit length uniquely specifies the vector. The solutions are $v^1 = -0.4991v^3$, $v^2 = 0.0701v^3$, Vector $\mathbf{v}$ will have unit length if $v^3 = 0.0928$. Hence, the eigenvector associated with $\lambda = 0.4817$ is $\mathbf{v} = [-0.0463, 0.0065, 0.0928]$. Computation of the other eigenvectors is the subject of Exercise 3–33; the reader might wish to carry out the whole calculation by means of Eqs. (3–121) and (3–117).

An important property of the eigenvectors corresponding to different (nondegenerate) eigenvalues of a symmetric second-rank tensor is that they are orthogonal. To prove this, suppose to be specific that $\mathbf{T}$ is doubly contravariant, λ and η are two eigenvalues of $\mathbf{T}$, and $\mathbf{u}$ and $\mathbf{v}$ are the corresponding eigenvectors. Then

$$T^{ij}u_j = \lambda u^i \tag{3–122}$$

We multiply both sides of Eq. (3–122) by v_i and sum over i.

$$v_iT^{ij}u_j = \lambda u^iv_i \tag{3–123}$$

Similarly, we construct

$$u_iT^{ij}v_j = \eta v^iu_i \tag{3–124}$$

Subtracting Eq. (3–124) from Eq. (3–123),

$$v_iT^{ij}u_j - u_iT^{ij}v_j = (\lambda - \eta)u^iv_i \tag{3–125}$$

Since $\mathbf{T}$ is symmetric, we may replace T^{ij} in one of the terms of Eq. (3–125) by T^{ji}. Thus,

$$v_iT^{ji}u_j - u_iT^{ij}v_j = (\lambda - \eta)u^iv_i \tag{3–126}$$

We are free to exchange the dummy indices i and j in the first term in this equation, and this shows that the entire left-hand side vanishes. Since we have stipulated that $\lambda \neq \eta$, it follows that u^iv_i must vanish, which proves that $\mathbf{u}$ and $\mathbf{v}$ are orthogonal.

The next property of eigenvectors that we shall examine is their utility in deducing a transformation matrix that will transform a symmetric tensor **T** to a cartesian coordinate system in which the axes are the eigenvectors of **T**. Let $\mathbf{v}(I)$ be the eigenvector corresponding to eigenvalue $\lambda(I)$, and let **V** be the matrix for which column I is $\mathbf{v}(I)$. That is,

$$\mathbf{V} = \begin{pmatrix} v^1(1) & v^1(2) & v^1(3) \\ v^2(1) & v^2(2) & v^2(3) \\ v^3(1) & v^3(2) & v^3(3) \end{pmatrix} \tag{3–127}$$

By Eq. (3–116) and the rules for matrix multiplication

$$\mathbf{TgV} = \mathbf{V} \begin{pmatrix} \lambda(1) & 0 & 0 \\ 0 & \lambda(2) & 0 \\ 0 & 0 & \lambda(3) \end{pmatrix} \tag{3–128}$$

Multiplying both sides of Eq. (3–128) on the left by $\bar{\mathbf{V}}\mathbf{g}$ and using the consequence of the orthonormality of the eigenvectors that

$$\bar{\mathbf{V}}\mathbf{gV} = \mathbf{I} \tag{3–129}$$

we obtain

$$\bar{\mathbf{V}}\mathbf{gTgV} = \begin{pmatrix} \lambda(1) & 0 & 0 \\ 0 & \lambda(2) & 0 \\ 0 & 0 & \lambda(3) \end{pmatrix} \tag{3–130}$$

The matrix

$$\mathbf{F} = \bar{\mathbf{V}}\mathbf{g} \tag{3–131}$$

thus constitutes a transformation matrix that diagonalizes **T** by transforming to its principal axes. The eigenvalues $\lambda(I)$ then give the magnitudes of **T** along its principal axes. In the case of anisotropic temperature factors, for example,

$$\lambda(I) = 2\pi^2\, \overline{(u(I))^2} \tag{3–132}$$

where $\overline{(\mathrm{u}(I))^2}$ is the mean-square displacement of the atom along the direction of principal axis I of β.

Exercise 3–17 showed that, for a second-rank contravariant tensor **T**, the product $T^{ij}g_{ij}$ is invariant upon transformation and is a generalization of the invariance of the trace in cartesian systems. The product $T^{ij}g_{ij}$ is in fact the trace of the matrix **Tg**, so we may say that the trace of **Tg** is invariant. If the eigenvalues $\lambda(I)$ of **T** have been found, the trace of the diagonalized matrix is just $\lambda(1) + \lambda(2) + \lambda(3)$. If this result is applied to anisotropic temperature factors, we have a means of obtaining a mean isotropic B corresponding to a set of β^{ij}. By Eqs. (3–112) and (3–132),

$$\overline{B} = 4\ \mathrm{trace}(\beta\mathbf{g})/3 \qquad (3\text{–}133)$$

(See Hamilton, 1959.)

EXERCISE 3–33 Calculate the eigenvectors associated with the other two eigenvalues (0.3500 and 0.1481) of the example of this section. Confirm that the three eigenvectors are orthonormal.

EXERCISE 3–34 Calculate the matrix $\mathbf{F} = \overline{\mathbf{V}}\mathbf{g}$ for the example of this section and compute $\mathbf{F}\beta\overline{\mathbf{F}}$.

EXERCISE 3–35 What are the eigenvalues and eigenvectors of the metric tensor **g**? Discuss the significance of your answer.

EXERCISE 3–36 Calculate the eigenvalues, eigenvectors, principal root-mean-square displacements, transformation matrix $\mathbf{F} = \overline{\mathbf{V}}\mathbf{g}$, and $\mathbf{F}\beta\overline{\mathbf{F}}$ for the triclinic anisotropic temperature factors of Exercise 3–30.

EXERCISE 3–37 Calculate the matrix **G**, which is the contragredient of the matrix **F** obtained in Exercise 3–36, and calculate the metric tensor of the coordinate system produced by this transformation by using Eq. (3–32).

EXERCISE 3–38 The crystal structure of 3,4-epoxysulfolane [*Acta Cryst.* **B28**, 2463 (1972)] is orthorhombic with $a_1 = 8.475$, $a_2 = 10.742$, $a_3 = 5.899$ Å. The anisotropic temperature factor of the sulfur atom has components $\beta^{11} = 0.00906$, $\beta^{12} = -0.00049$, $\beta^{13} = -0.00102$, $\beta^{22} = 0.00401$, $\beta^{23} = 0.00038$, $\beta^{33} = 0.01424$. Find the eigenvalues and eigenvectors of β and the root-mean-square displacements along the principal axes.

EXERCISE 3–39 Calculate the trace of $\beta\mathbf{g}$ for the temperature factors of Exercise 3–30 and compare with the sum of the eigenvalues that were computed in Exercise 3–36.

EXERCISE 3–40 Use Eq. (3–133) to calculate an average isotropic temperature factor

$\bar{B}$ corresponding to the temperature factors of Exercise 3–30 (see Exercises 3–36 and 3–39).

EXERCISE 3–41 Use Eq. (3–133) to calculate an average isotropic temperature factor $\bar{B}$ corresponding to the temperature factors of the sulfur atom in Exercise 3–38.

EXERCISE 3–42 Figure 3–1 shows the representation surface for the two-dimensional cartesian tensor $\mathbf{T} = \begin{pmatrix} 1.4 & 0.4 \\ 0.4 & 1.0 \end{pmatrix}$. The principal axes are $\mathbf{v}(1)$ and $\mathbf{v}(2)$. Vector $\mathbf{s}$ is produced by the operation $\mathbf{Tr}$. If $\mathbf{r} = [\sqrt{5}/4, -\sqrt{5}/5]$, calculate the angle between $\mathbf{r}$ and $\mathbf{s}$.

EXERCISE 3–43 If the doubly contravariant tensor $\mathbf{T}$ operates on vector $\mathbf{u}$ to produce vector $\mathbf{v}$, show that the scalar product of $\mathbf{u}$ and $\mathbf{v}$ is $T^{ij}u_i u_j$.

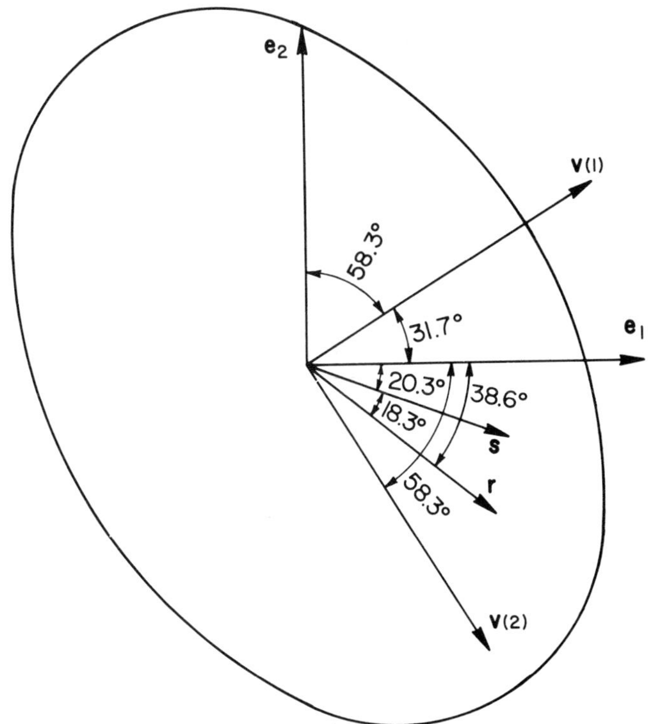

FIGURE 3–1. Representation surface for two-dimensional cartesian tensor $\mathbf{T} = \begin{pmatrix} 1.4 & 0.4 \\ 0.4 & 1.0 \end{pmatrix}$. Principal axes are $\mathbf{v}(1)$ and $\mathbf{v}(2)$. $\mathbf{T}$ operates on $\mathbf{r}$ to produce $\mathbf{s}$.

EXERCISE 3–44 Show that a nonsymmetric doubly contravariant or covariant second-rank tensor cannot be diagonalized by a linear transformation.

3–15 Magnitude of Second-Rank Tensor

If **T** is a second-rank tensor, and **u** is a unit vector, the operation **Tu** produces a new vector **v**; that is,

$$\mathbf{v} = \mathbf{Tu} \tag{3–134}$$

The magnitude $T(\mathbf{u})$ of the tensor **T** in the direction of the unit vector **u** is equal to the length of the projection of **v** onto **u**. That is,

$$T(\mathbf{u}) = \mathbf{u} \cdot \mathbf{v} = \bar{\mathbf{u}}\mathbf{Tu} \tag{3–135}$$

In terms of components

$$T(\mathbf{u}) = T^{ij}u_i u_j = T_{ij}\, u^i u^j \tag{3–136}$$

If **u** is an eigenvector of **T**, T(**u**) is of course just the corresponding eigenvalue. A physical interpretation of the magnitude of a second-rank tensor will be described in Section 5–2.

EXERCISE 3–45 Derive a formula for the magnitude in the direction of reciprocal axis $\mathbf{a}^k$ of an anisotropic temperature factor tensor with components β^{ij}.

EXERCISE 3–46 Calculate the magnitude of the anisotropic temperature factor tensor of Exercise 3–38 in the crystal directions (a) [1, 0, 0]; (b) [0, 1, 0]; (c) [0, 0 ,1]; (d) [−0.1097, 0.0210, 0.0492].

3–16 Rigid-Body Motion

The thermal motion of atoms in a molecular crystal often may be resolved into the intermolecular movements of the molecules as units and the intramolecular movements of the atoms within the molecules relative to each other. Separation of the rigid-body molecular motion from the total movement is particularly clearcut if the amplitudes of molecular motion are large compared with the vibrational amplitudes within the molecules. Determination of the rigid-body motion from the observed anisotropic temperature factors was first described by Cruickshank (1956). A rigorous theory, which permitted nonintersecting axes of libration, was developed by Schomaker and Trueblood (1968), and this is the theory that will be described in this section. The original treatment of Schomaker and Trueblood used the dyadic notation, which is outlined in Section 3-17. A detailed and lucid matrix formulation has been presented by Johnson (1970) and Johnson and Levy (1974), and their treatment will be followed here; great emphasis will be placed here, however, upon analysis of rigid-body motion as an example of the application of tensor methods to problems in generalized coordinate systems.

The general displacement of a molecule from its equilibrium position and orientation consists of a translation, given by vector **t**, and a rotation. The operators for finite rotations do not commute, and consequently finite rotations cannot be represented by vectors. An adequate treatment of anisotropic thermal motion, accurate to the level of the quadratic approximation, can be based upon infinitesimal rotations, which do commute (cf. Goldstein, 1980). To illustrate this vectorial representation of rotation, consider an atom at $[r^1, r^2, r^3]$ in a cartesian coordinate system. Rotation through angle θ about basis vector $\mathbf{e}_3$ moves it to $[r^1 \cos\theta - r^2 \sin\theta, r^1 \sin\theta + r^2 \cos\theta, r^3]$. If θ is small, the change in position may be approximated as $[-r^2\theta, r^1\theta, 0]$, and this is equal to $\boldsymbol{\lambda} \wedge \mathbf{r}$, where $\boldsymbol{\lambda}$ is the vector $[0, 0, \theta]$. The form $\boldsymbol{\lambda} \wedge \mathbf{r}$, where $\boldsymbol{\lambda}$ is a vector of length θ along the axis of rotation, is the general displacement due to the rotation.

The total displacement of an atom in a molecular crystal from its equilibrium position is

$$\mathbf{u} = \mathbf{t} + \boldsymbol{\lambda} \wedge \mathbf{r} \tag{3-137}$$

or, in suffix notation,

$$u^i = t^i + \epsilon_{klm} \lambda^l r^m g^{ki} \tag{3-138}$$

The average values of all products $u^i u^j$ constitute the components of the doubly contravariant tensor **U**:

$$U^{ij} = \langle u^i u^j \rangle \tag{3-139}$$

$$\begin{aligned} U^{ij} = {} & \langle t^i t^j \rangle + \epsilon_{klm}\epsilon_{nop} r^m r^p g^{ki} g^{nj} \langle \lambda^l \lambda^o \rangle \\ & + \epsilon_{nop} r^p g^{nj} \langle \lambda^o t^i \rangle + \epsilon_{klm} r^m g^{ki} \langle \lambda^l t^j \rangle \end{aligned} \tag{3-140}$$

The relationship between U^{ij} and β^{ij} is given by Eq. (3–105). The various averages in Eq. (3–140) may be identified as a translation tensor **T** with components

$$T^{ij} = <t^i t^j> \tag{3–141}$$

a libration tensor **L** with components

$$L^{ij} = <\lambda^i \lambda^j> \tag{3–142}$$

and a screw tensor **S** with components*

$$S^{ij} = <\lambda^i t^j> \tag{3–143}$$

Defining $A^i{}_l$ by

$$A^i{}_l = \epsilon_{klm} r^m g^{ki} \tag{3–144}$$

Eq. (3–140) may be written

$$U^{ij} = T^{ij} + A^i{}_l\, A^j{}_o\, L^{lo} + A^j{}_o\, S^{oi} + A^i{}_l\, S^{lj} \tag{3–145}$$

which in matrix notation is

$$\mathbf{U} = \mathbf{T} + \mathbf{AL\bar{A}} + \mathbf{\bar{S}\bar{A}} + \mathbf{AS} \tag{3–146}$$

Following Johnson (1970), this may be expressed as a partitioned matrix equation

$$\mathbf{U} = \left[\mathbf{I} \middle| \mathbf{A}\right] \left[\begin{array}{c|c} \mathbf{T} & \mathbf{\bar{S}} \\ \hline \mathbf{S} & \mathbf{L} \end{array}\right] \left[\begin{array}{c} \mathbf{I} \\ \hline \mathbf{\bar{A}} \end{array}\right] \tag{3–147}$$

Equation (3–147) follows the usual rules for multiplication of a 1 × 2 matrix by a 2 × 2 matrix by a 2 × 1 matrix, except that the individ-

*The symbol **S** is used in Chapter 4 to represent a general symmetry operator. The meaning will always be apparent from the context, so no confusion should result from following conventional usage even though the same symbol is used for two different things.

ual terms are themselves matrices whose products are given by matrix multiplication (see, for example, Ayres, 1962). The tensors **U, T, L,** and **S** are all second-rank contravariant tensors with transformation properties

$$\mathbf{U}' = \mathbf{F}\mathbf{U}\bar{\mathbf{F}} \tag{3–148}$$

$$\mathbf{T}' = \mathbf{F}\mathbf{T}\bar{\mathbf{F}} \tag{3–149}$$

$$\mathbf{L}' = \mathbf{F}\mathbf{L}\bar{\mathbf{F}} \tag{3–150}$$

$$\mathbf{S}' = \mathbf{F}\mathbf{S}\bar{\mathbf{F}} \tag{3–151}$$

The transformation properties of **A** follow from Eq. (3–146):

$$\mathbf{A}' = \mathbf{F}\mathbf{A}\mathbf{F}^{-1} = \mathbf{F}\mathbf{A}\bar{\mathbf{G}} \tag{3–152}$$

Shifting the origin by vector ϱ converts positional vector **r** into $\mathbf{r} - \varrho$. The rotational vector λ and the atomic displacements from equilibrium **u** are not affected by this translation. Thus, translation produces

$$\mathbf{u}' = \mathbf{u} = \mathbf{t}' + \lambda \wedge (\mathbf{r} - \varrho) \tag{3–153}$$

Hence, by Eq. (3–137)

$$\mathbf{t}' = \mathbf{t} + \lambda \wedge \varrho \tag{3–154}$$

$$t'^{i} = t^{i} + \epsilon_{klm} \lambda^{l} \varrho^{m} g^{ki} \tag{3–155}$$

Multiplying by t'^{j} and averaging produces

$$T'^{ij} = T^{ij} + \epsilon_{klm}\epsilon_{nop}\varrho^{m}\varrho^{p}g^{ki}g^{nj}L^{lo} + \epsilon_{nop}\varrho^{p}g^{nj}S^{oi} + \epsilon_{klm}\varrho^{m}g^{ki}S^{lj} \tag{3–156}$$

A matrix **P** analogous to **A** may be defined in terms of the shifts

$$P^{i}{}_{l} = \epsilon_{klm}\varrho^{m}g^{ki} \tag{3–157}$$

$$T'^{ij} = T^{ij} + P^{i}{}_{l}\, P^{j}{}_{o}\, L^{lo} + P^{j}{}_{o}\, S^{oi} + P^{i}{}_{l}\, S^{lj} \tag{3-158}$$

$$\mathbf{T}' = \mathbf{T} + \mathbf{PL\bar{P}} + \mathbf{\bar{S}\bar{P}} + \mathbf{PS} \tag{3-159}$$

The shift of origin transforms the screw tensor according to

$$S'^{ij} = \langle \lambda'^{i} t'^{j} \rangle = \langle \lambda^{i} (t^{j} + \epsilon_{klm}\, \lambda^{l}\, \varrho^{m} g^{kj}) \rangle \tag{3-160}$$

$$S'^{ij} = S^{ij} + P^{j}{}_{l}\, L^{li} \tag{3-161}$$

$$\mathbf{S}' = \mathbf{S} + \mathbf{\bar{L}\bar{P}} \tag{3-162}$$

The effect of the origin shift on **A** is obtained by replacing **r** by **r** $-\varrho$. Hence,

$$\mathbf{A}' = \mathbf{A} - \mathbf{P} \tag{3-163}$$

The total transformation properties of these tensors, under change of axes followed by shift of origin, are

$$\mathbf{U}' = \mathbf{FU\bar{F}} \tag{3-164}$$

$$\mathbf{L}' = \mathbf{FL\bar{F}} \tag{3-165}$$

$$\mathbf{T}' = \mathbf{FT\bar{F}} + \mathbf{PFL\bar{F}\bar{P}} + \mathbf{F\bar{S}\bar{F}\bar{P}} + \mathbf{PFS\bar{F}} \tag{3-166}$$

$$\mathbf{S}' = \mathbf{FS\bar{F}} + \mathbf{FL\bar{F}\bar{P}} \tag{3-167}$$

$$\mathbf{A}' = \mathbf{FA\bar{G}} - \mathbf{P} \tag{3-168}$$

The components of **T**, **L**, and **S** may be determined by carrying out a least-squares fit of the observed U^{ij} to Eq. (3–145). Since **T** and **L** are symmetric, with six independent components each, and **S** is unsymmetric and has nine independent components, there are ostensibly 21 unknowns to be determined from the least-squares treatment. However, the system of equations possesses a redundancy that limits the number of elements of **S** that can be determined to eight. The redundancy arises because there exist matrices **D** such that **S** + **D** makes the same contribution to **U** as **S** itself. The condition that **D** must satisfy is found from Eq. (3–146) to be

$$\bar{\mathbf{D}}\bar{\mathbf{A}} + \mathbf{A}\mathbf{D} = \mathbf{0} \tag{3–169}$$

In a cartesian coordinate system $\bar{\mathbf{A}}$ is equal to $-\mathbf{A}$ (see Table 3–1), and a solution of Eq. (3–169) is

$$\mathbf{D} = k\mathbf{I} \tag{3–170}$$

where k is an arbitrary constant. Since **D** must transform in the same way as **S**, transforming to crystal axes produces

$$\mathbf{D} = \mathbf{F}k\mathbf{I}\bar{\mathbf{F}} = k\mathbf{g}^* \tag{3–171}$$

Thus, the values of U^{ij} computed from the components of **T**, **L**, and **S** are unchanged if a multiple of **g*** is added to **S**. In cartesian coordinates the condition is that only the differences $S^{11} - S^{22}$, $S^{22} - S^{33}$, and $S^{33} - S^{11}$ may be obtained from the measured U^{ij}, and, since the sum of two of these quantities equals the negative of the third, the number of determinable components is reduced by one. This limitation was recognized in the original paper of Schomaker and Trueblood (1968) and has been discussed by Pawley (1968) and others. Pawley, in a private communication, has pointed out that all the S^{ij} components are determined uniquely by lattice dynamics.

Analysis of the rigid-body molecular motion is most easily carried out in a cartesian coordinate system, in which the indeterminancy in **S** affects only the diagonal elements, and the indeterminancy may be handled by constraining the trace of **S** to a particular value, say zero. The 20 determinable elements of the tensors are obtained by a least-

TABLE 3–1 COMPONENTS OF $A^i{}_j$ DERIVED FROM EQ. (3–144)

$$\begin{aligned}
A^1{}_1 &= PV(r^2g^{31} - r^3g^{21})\\
A^1{}_2 &= PV(r^3g^{11} - r^1g^{31})\\
A^1{}_3 &= PV(r^1g^{21} - r^2g^{11})\\
A^2{}_1 &= PV(r^2g^{32} - r^3g^{22})\\
A^2{}_2 &= PV(r^3g^{12} - r^1g^{32})\\
A^2{}_3 &= PV(r^1g^{22} - r^2g^{12})\\
A^3{}_1 &= PV(r^2g^{33} - r^3g^{23})\\
A^3{}_2 &= PV(r^3g^{13} - r^1g^{33})\\
A^3{}_3 &= PV(r^1g^{23} - r^2g^{13})
\end{aligned}$$

squares fit of the observed U^{ij} to Eq. (3–145). Schomaker and Trueblood recommend then transforming to the principal axes of **L** and shifting the origin so as to make **S** symmetric; this shift minimizes the trace of **T**. Details of the analysis, including the determination of three nonintersecting axes of libration and the pitches of the screws coupling the translations with the librations, are given in the works of Schomaker and Trueblood (1968), Johnson, (1970), and Johnson and Levy (1974).

For illustration, Table 3–2 lists the values of U^{ij} for the atoms other than hydrogen in a crystal of $C_4H_6OSO_2$ [*Acta Cryst.* **B28**, 2463–2468 (1972)]. The values given are referred to a cartesian coordinate system. The calculated values are based upon the tensor components in Table 3–3, which were determined from the measured values by the least-squares process.

TABLE 3–2 CARTESIAN PARAMETERS OF ATOMS IN $C_4H_6OSO_2$

Atom	Cartesian coordinates	ij	U^{ij} (obs)	U^{ij} (calc)
S	1.1378 1.2042, 0.9568	11	0.0330	0.0366
		12	−0.0023	−0.0022
		13	−0.0026	−0.0022
		22	0.0234	0.0234
		23	0.0012	0.0013
		33	0.0251	0.0251
O(1)	2.548, 0.912, 0.838	11	0.0290	0.0294
		12	−0.0044	−0.0048
		13	−0.0094	−0.0110
		22	0.0489	0.0491
		23	0.0099	0.0098
		33	0.0544	0.0558
O(2)	0.711, 2.000, 2.080	11	0.0876	0.0775
		12	0.0006	−0.0018
		13	0.0017	−0.0021
		22	0.0365	0.0359
		23	−0.0060	−0.0087
		33	0.0269	0.0270
O(3)	−1.697, 0.999, 0.025	11	0.0248	0.0231
		12	0.0071	0.0059
		13	0.0068	0.0068
		22	0.0496	0.0535
		23	0.0067	0.0067

TABLE 3-2 (*Continued*)

Atom	Cartesian coordinates	ij	U^{ij} (obs)	U^{ij} (calc)
		33	0.0514	0.0525
C(1)	0.198, −0.325, 0.892	11	0.0360	0.0342
		12	−0.0041	−0.0020
		13	−0.0002	0.0004
		22	0.0245	0.0227
		23	0.0044	0.0033
		33	0.0368	0.0366
C(2)	0.521, 1.930, −0.563	11	0.0341	0.0323
		12	−0.0022	−0.0021
		13	0.0020	0.0007
		22	0.0310	0.0312
		23	0.0073	0.0064
		33	0.0269	0.0260
C(3)	−0.822, −0.127, −0.186	11	0.0306	0.0285
		12	−0.0066	−0.0056
		13	0.0016	−0.0004
		22	0.0346	0.0338
		23	−0.0037	−0.0013
		33	0.0382	0.0398
C(4)	−0.656, 1.086, −0.964	11	0.0269	0.0276
		12	0.0005	0.0001
		13	−0.0025	−0.0019
		22	0.0412	0.0407
		23	0.0021	0.0021
		33	0.0306	0.0294

EXERCISE 3–47 Show that

$$\mathbf{A} = \bar{\mathbf{g}}^* \begin{pmatrix} 0 & r^3 & -r^2 \\ -r^3 & 0 & r^1 \\ r^2 & -r^1 & 0 \end{pmatrix} VP$$

where *P is the parity operator (*see Sections 1–14 and 3–11).

EXERCISE 3–48 Derive the transformation properties of **A** given by Eq. (3–152) by inserting into Eq. (3–144) the transformation properties of ϵ_{klm} [from Eq. (3–91)], of **r**, and of **g***.

EXERCISE 3–49 Insert Eqs. (3–165), (3–166), (3–167), and (3–168) into **T**′

$+ \mathbf{A}'\mathbf{L}'\overline{\mathbf{A}}' + \overline{\mathbf{S}}'\overline{\mathbf{A}}' + \mathbf{A}'\mathbf{S}'$, and verify that these transformations are consistent with Eq. (3-164).

EXERCISE 3-50 Using the tensor components in Table 3-3, calculate the components of U^{ij} for a sulfur atom with cartesian coordinates [1.1378 1.2042, 0.9568]. (Use $S^{11} = 0.00466$, $S^{22} = 0.00392$, $S^{33} = 0$.) Compare your answers with the values in Table 3-2.

EXERCISE 3-51 Repeat Exercise 3-49 using the values $S^{11} = 0.01466$, $S^{22} = 0.01392$, $S^{33} = 0.01000$, thus confirming that only the differences in the S^{ii} contribute to **U** in a cartesian system.

TABLE 3-3 RIGID-BODY TENSOR COMPONENTS IN $C_4H_6OSO_2$

$L^{11} = 0.00747$	$T^{11} = 0.02639$
$L^{12} = 0.00137$	$T^{12} = 0.00001$
$L^{13} = 0.00242$	$T^{13} = 0.00233$
$L^{22} = 0.00716$	$T^{22} = 0.02415$
$L^{23} = -0.00177$	$T^{23} = -0.00417$
$L^{33} = 0.00737$	$T^{33} = 0.03081$
$S^{11} - S^{22} = 0.00074$	$S^{22} - S^{33} = 0.0392$
$S^{12} = 0.00330$	$S^{21} = -0.00064$
$S^{13} = -0.00786$	$S^{31} = 0.00536$
$S^{23} = 0.00139$	$S^{32} = -0.00240$

3-17 Dyadics

A dyad is constructed from a pair of vectors written side by side, as **uv**, with no operation defined between the vectors. A dyad operates upon another vector according to the definitions

$$\mathbf{uv} \cdot \mathbf{w} = \mathbf{u}(\mathbf{v} \cdot \mathbf{w}) \tag{3-172}$$

$$\mathbf{w} \cdot \mathbf{uv} = (\mathbf{w} \cdot \mathbf{u})\mathbf{v} \tag{3-173}$$

A dyadic is a sum of dyads; for example

$$\mathbf{\Phi} = \mathbf{qr} + \mathbf{st} + \mathbf{uv} \tag{3-174}$$

with the operational properties

$$\mathbf{\Phi} \cdot \mathbf{w} = \mathbf{q}(\mathbf{r} \cdot \mathbf{w}) + \mathbf{s}(\mathbf{t} \cdot \mathbf{w}) + \mathbf{u}(\mathbf{v} \cdot \mathbf{w}) \tag{3-175}$$

$$\mathbf{w} \cdot \mathbf{\Phi} = (\mathbf{w} \cdot \mathbf{q})\mathbf{r} + (\mathbf{w} \cdot \mathbf{s})\mathbf{t} + (\mathbf{w} \cdot \mathbf{u})\mathbf{v} \tag{3-176}$$

A dyadic is thus similar to a tensor in that it operates upon a vector to produce another vector. Specifically, dyadics have properties comparable to those of second-rank tensors, and they provide an alternative to the tensor formulations that are the main topic of this book. Tensors have advantages in computation, in unambiguous notation, and in generalization; dyadics, on the other hand, often lead to compact expression of certain relationships. Operators analogous to higher rank tensors can be constructed (triadics, etc.), but tensor methods make such quantities unnecessary.

Zachariasen (1945) used dyadics to derive the crystallographic space groups, and his book includes an excellent summary of dyadic algebra. Patterson (1959) has given a concise but lucid treatment of dyadics. The present treatment follows the notation of Zachariasen and Patterson. As was mentioned in Section 3–16, Shomaker and Trueblood (1968) based their analysis of rigid-body motion upon dyadics, and their article exemplifies application to a complex problem.

The following properties of dyadics are intended to serve an an introduction and an illustration of their essential features.

The dyadic conjugate to that of Eq. (3–174) is

$$\mathbf{\Phi}_c = \mathbf{rq} + \mathbf{ts} + \mathbf{vu} \tag{3–177}$$

The identity operator, which is known as the idemfactor, is

$$\mathbf{I} = \mathbf{a}^i\mathbf{a}_i = \mathbf{a}_i\mathbf{a}^i \tag{3–178}$$

The most general form of a dyadic in three-dimensional space consists of nine terms

$$\mathbf{\Phi} = \phi^{ij}\mathbf{a}_i\mathbf{a}_j = \phi_{ij}\mathbf{a}^i\mathbf{a}^j = \phi^i{}_j\,\mathbf{a}_i\mathbf{a}^j = \phi_i{}^j\,\mathbf{a}^i\mathbf{a}_j \tag{3–179}$$

where the ϕ^{ij}, ϕ_{ij}, $\phi^i{}_j$, and $\phi_i{}^j$ are numerical coefficients; this is referred to as the nonion form. These numerical coefficients are in fact tensor components.

The scalar of a dyadic is obtained by taking the scalar product of the vectors in each dyad; for the dyadic of Eq. (3–174) the scalar is

$$\mathbf{\Phi}_s = \mathbf{q}\cdot\mathbf{r} + \mathbf{s}\cdot\mathbf{t} + \mathbf{u}\cdot\mathbf{v} \tag{3–180}$$

Similarly, the vector of a dyadic is obtained by taking the vector product within each dyad, and the vector of Eq. (3–174) is

$$\boldsymbol{\Phi}_v = \mathbf{q} \wedge \mathbf{r} + \mathbf{s} \wedge \mathbf{t} + \mathbf{u} \wedge \mathbf{v} \tag{3–181}$$

If $\boldsymbol{\Phi}$ is the dyadic of Eq. (3–174) and $\boldsymbol{\Psi}$ is the dyadic

$$\boldsymbol{\Psi} = \mathbf{kl} + \mathbf{mn} + \mathbf{op} \tag{3–182}$$

the scalar products of $\boldsymbol{\Phi}$ and $\boldsymbol{\Psi}$ are defined as

$$\boldsymbol{\Phi} \cdot \boldsymbol{\Psi} = (\mathbf{r} \cdot \mathbf{k})\mathbf{ql} + (\mathbf{t} \cdot \mathbf{m})\mathbf{sn} + (\mathbf{v} \cdot \mathbf{o})\mathbf{up} \tag{3–183}$$

$$\boldsymbol{\Psi} \cdot \boldsymbol{\Phi} = (\mathbf{l} \cdot \mathbf{q})\mathbf{kr} + (\mathbf{n} \cdot \mathbf{s})\mathbf{mt} + (\mathbf{p} \cdot \mathbf{u})\mathbf{ov} \tag{3–184}$$

The determinant of dyadic $\boldsymbol{\Phi}$ of Eq. (3–174) is

$$|\boldsymbol{\Phi}| = (\mathbf{q} \cdot \mathbf{s} \wedge \mathbf{u})\,(\mathbf{r} \cdot \mathbf{t} \wedge \mathbf{v}) \tag{3–185}$$

That is, the determinant is given by the volume of the parallelepiped spanned by the prefactors multiplied by the volume of the parallelepiped spanned by the postfactors. In terms of the nonion form,

$$|\boldsymbol{\Phi}| = g|\phi^{ij}| = g^*|\phi_{ij}| = |\phi^{i}{}_{j}| = |\phi_{i}{}^{j}| \tag{3–186}$$

If $|\boldsymbol{\Phi}|$ is 0, the dyadic is said to be singular or incomplete.
If $|\boldsymbol{\Phi}|$ is nonzero, the reciprocal dyadic is defined by

$$\boldsymbol{\Phi} \cdot \boldsymbol{\Phi}^{-1} = \boldsymbol{\Phi}^{-1} \cdot \boldsymbol{\Phi} = \mathbf{I} \tag{3–187}$$

In Section 4–7 we shall examine the representation of symmetry operators by dyadics that possess the property

$$\boldsymbol{\Phi} = \boldsymbol{\Phi}_C^{-1} \tag{3–188}$$

Dyadics with this property are called versors.

The transformation properties of dyadics correspond, of course, to those of second-rank tensors, which may be shown by requiring that $\mathbf{h} \cdot \boldsymbol{\Phi} \cdot \mathbf{h}$ be invariant. Thus,

$$\phi'^{ij} = F^{i}{}_{k}\, F^{j}{}_{l}\, \phi^{kl} \tag{3–189}$$

Exercise 3–52 If $\mathbf{x} = \mathbf{w} \cdot \mathbf{\Phi}$ and $\mathbf{w} = \mathbf{z} \cdot \mathbf{\Psi}$, show that $\mathbf{x} = \mathbf{z} \cdot \mathbf{\Psi} \cdot \mathbf{\Phi}$.

EXERCISE 3–53 Show that $\mathbf{r} \cdot \mathbf{I} = \mathbf{I} \cdot \mathbf{r} = \mathbf{r}$, where $\mathbf{r}$ is any vector and $\mathbf{I}$ is given by Eq. (3–178).

EXERCISE 3–54 Write the nonion forms of **I.**

EXERCISE 3–55 Calculate the scalar of the nonion form of Eq. (3–179).

EXERCISE 3–56 Express ϕ^{ij} in terms of the components of **u** and **v** for the dyad **uv**.

EXERCISE 3–57 A dyadic is said to be symmetric if $\mathbf{\Phi} = \mathbf{\Phi}_C$. Show that the vector of a symmetric dyadic is zero.

EXERCISE 3–58 A dyadic is said to be antisymmetric if $\mathbf{\Phi} = -\mathbf{\Phi}_C$. Show that both the scalar and the determinant of an antisymmetric dyadic vanish.

EXECISE 3–59 Show that for any dyadic

$$\mathbf{I} \cdot \mathbf{\Phi} = \mathbf{\Phi} \cdot \mathbf{I} = \mathbf{\Phi}$$

EXERCISE 3–60 Derive Eq. (3–186) by applying Eq. (3–185) to the nonion forms of Eq. (3–179).

EXERCISE 3–61 Show that for any two dyadics $\mathbf{\Phi}$ and $\mathbf{\Psi}$, $(\mathbf{\Phi} \cdot \mathbf{\Psi})^{-1} = \mathbf{\Psi}^{-1} \cdot \mathbf{\Phi}^{-1}$, and $(\mathbf{\Phi} \cdot \mathbf{\Psi})_C = \mathbf{\Psi}_C \cdot \mathbf{\Phi}_C$.

EXERCISE 3–62 Show that the determinant of a versor is ± 1.

EXERCISE 3–63 Express the temperature factor of Exercise 3–38 as a dyadic in nonion form.

EXERCISE 3–64 The nonion form of a symmetric dyadic has components $\phi^{11} = 0.2400$, $\phi^{12} = 0.0200$, $\phi^{13} = -0.0100$, $\phi^{22} = 0.3600$, $= 0.1000$, $\phi^{33} = 0.3200$. Calculate the components of the dyadic in a new coordinate system chosen by the transformation of Exercise 3–4.

EXERCISE 3–65 Show that

$$\mathbf{I} \wedge \mathbf{u} = \epsilon_{ijk} u^j \mathbf{a}^i \mathbf{a}^k$$

3–18 Summary of Transformations

Table 3–4 summarizes the transformation properties that have been developed in this chapter. Included for completeness are the transformation properties of point symmetry operators, which will be shown in Chapter 4 to be mixed seond-rank tensors, with one contravariant and one covariant index.

TABLE 3-4 SUMMARY OF TRANSFORMATIONS

	Matrix notation	Tensor notation
Basis vectors		
	$\mathbf{a}' = \mathbf{Ga}$	$\mathbf{a}_i' = G_i{}^j\,\mathbf{a}_j$
	$\mathbf{F} = \overline{\mathbf{G}}^{-1}$	$F^i{}_j\,G_i{}^k = \delta_j{}^k$
	$\mathbf{a} = \overline{\mathbf{F}}\mathbf{a}'$	$\mathbf{a}_i = F^j{}_i\,\mathbf{a}_j'$
Coordinates		
	$\mathbf{x}' = \mathbf{Fx}$	$x'^i = F^i{}_j\,x^j$
	$\mathbf{x} = \overline{\mathbf{G}}\mathbf{x}'$	$x^i = G_j{}^i\,x'^j$
Covariant vector components		
	$\mathbf{h}' = \mathbf{Gh}$	$h_i' = G_i{}^j\,h_j$
	$\mathbf{h} = \overline{\mathbf{F}}\mathbf{h}'$	$h_i = F^j{}_i\,h_j'$
Metric tensor $g_{ij} = \mathbf{a}_i \cdot \mathbf{a}_j$		
	$\mathbf{g}' = \mathbf{Gg}\overline{\mathbf{G}}$	$g_{ij}' = G_i{}^k\,G_j{}^l\,g_{kl}$
	$\mathbf{g} = \overline{\mathbf{F}}\mathbf{g}'\mathbf{F}$	$g_{ij} = F^k{}_i F^l{}_j g_{kl}'$
Reciprocal basis vectors $\mathbf{a}_i \cdot \mathbf{a}^j = \delta_i{}^j$, $\mathbf{a}^{*j} = \mathbf{a}^j$		
	$\mathbf{a}'^* = \mathbf{Fa}^*$	$\mathbf{a}'^i = F^i{}_j\,\mathbf{a}^j$
	$\mathbf{a}^* = \overline{\mathbf{G}}\mathbf{a}'^*$	$\mathbf{a}^i = G_j{}^i\,\mathbf{a}'^j$
Reciprocal metric tensor $g^{ij} = \mathbf{a}^i \cdot \mathbf{a}^j$		
	$\mathbf{g}'^* = \mathbf{Fg}^*\overline{\mathbf{F}}$	$g'^{ij} = F^i{}_k\,F^j{}_l\,g^{kl}$
	$\mathbf{g}^* = \overline{\mathbf{G}}\mathbf{g}'^*\mathbf{G}$	$g^{ij} = G_k{}^i G_l{}^j g'^{kl}$
Second-rank contravariant tensor		
	$\mathbf{U}' = \mathbf{FU}\overline{\mathbf{F}}$	$U'^{ij} = F^i{}_k\,F^j{}_l\,U^{kl}$
	$\mathbf{U} = \overline{\mathbf{G}}\mathbf{U}'\mathbf{G}$	$U^{ij} = G_k{}^i\,G_l{}^j\,U'^{kl}$
Second-rank covariant tensor		
	$\mathbf{V}' = \mathbf{GV}\overline{\mathbf{G}}$	$V_{ij}' = G_i{}^k\,G_j{}^l\,V_{kl}$
	$\mathbf{V} = \overline{\mathbf{F}}\mathbf{V}'\mathbf{F}$	$V_{ij} = F^k{}_i\,F^l{}_j\,V_{kl}'$
Symmetry operator in real space		
	$\mathbf{y} = \mathbf{Sx}$	$y^i = S^i{}_j\,x^j$
	$\mathbf{S}' = \mathbf{FS}\overline{\mathbf{G}}$	$S'^i{}_j = F^i{}_k\,G_j{}^l\,S^k{}_l$
	$\mathbf{S} = \overline{\mathbf{G}}\mathbf{S}'\mathbf{F}$	$S^i{}_j = G_k{}^i\,F^l{}_j\,S'^k{}_l$
Symmetry operator in reciprocal space		
	$\mathbf{k} = \mathbf{Qh}$	$k_i = Q_i{}^l\,h_l$
	$\mathbf{Q}' = \mathbf{GQ}\overline{\mathbf{F}}$	$Q'_j{}^i = G_j{}^k\,F^i{}_l\,Q_k{}^l$
	$\mathbf{Q} = \overline{\mathbf{F}}\mathbf{Q}'\mathbf{G}$	$Q_j{}^i = F^i{}_k\,G_j{}^l\,Q_l'^k$

TABLE 3–4 (*Continued*)

Matrix notation	Tensor notation
Relationship between **S** and **Q**	
$\mathbf{S} = \mathbf{g}^*\mathbf{Q}\mathbf{g}$	$S^i{}_j = g^{ik} g_{lj} Q_k{}^l$
$\mathbf{Q} = \mathbf{g}\mathbf{S}\mathbf{g}^*$	$Q_i{}^j = g_{ik} g^{jl} S^k{}_l$
General transformation	
$A'^{p_1 \cdots p_m}{}_{q^1 \cdots q^n} = \dfrac{\partial x'^{p_1}}{\partial x^{i_1}} \cdots \dfrac{\partial x'^{p_m}}{\partial x^{i_m}} \dfrac{\partial x^{j_1}}{\partial x'^{q_1}} \cdots \dfrac{\partial x^{j_n}}{\partial x'^{q_n}} A^{i_1 \cdots i_m}{}_{j_1 \cdots j_n}$	

CHAPTER 4

Symmetry

4-1 Introduction

This chapter will deal with certain mathematical operations that generate the coordinates of new points from the coordinates of old points. The crystal axes will be assumed to be fixed in length and in direction; the operations thus are defined relative to a constant set of basis vectors. The operations of this type that are of special interest are the *symmetry operations* that produce points whose physical environments are equivalent to those of the original points. Within an infinite crystal, the new points are indistinguishable from the old except for possible mirror images or inversions.

A further requirement of the symmetry operations is that they apply to all points in the crystal. If $\mathbf{x}$ is the vector to any point in the crystal, symmetry operator $\mathbf{S}$ produces an equivalent point. The operation may be denoted by

$$\mathbf{y} = \mathbf{S}\mathbf{x} \tag{4-1}$$

Having produced the point denoted by vector $\mathbf{y}$, the combination $\mathbf{S}\mathbf{y}$ must indicate another point (or the original point back again). Thus, $\mathbf{S}\mathbf{S}\mathbf{x}$ is also a vector to an equivalent point, so $\mathbf{S}\mathbf{S}$ is a symmetry operation too. A consequence of this condition is that the set of symmetry operators must form a *group*, which is defined as a set of elements for which there exists a law of combination (called multiplication) such that, if $\mathbf{Q}$, $\mathbf{R}$, and $\mathbf{S}$ are elements of the group,

(1) $\mathbf{Q}(\mathbf{R}\mathbf{S}) = (\mathbf{Q}\mathbf{R})\mathbf{S}$ (associative law) (4-2)

(2) there exists an identity element $\mathbf{I}$ which has the property

$$\mathbf{S}\mathbf{I} = \mathbf{I}\mathbf{S} = \mathbf{S} \tag{4-3}$$

for any element of the group;

(3) for any element $\mathbf{S}$ there is in the group an inverse element $\mathbf{S}^{-1}$ with the property

$$\mathbf{S}\mathbf{S}^{-1} = \mathbf{S}^{-1}\mathbf{S} = \mathbf{I} \tag{4-4}$$

(4) the product of any two elements is an element of the group.

The notational and classification systems for symmetry groups have been treated at length in many books (see, for example, *Introduction to Crystallography*, Chapters 2 and 4) and will not be repeated here. The concern here is rather in the tensor description of symmetry and in the development of the transformation properties of symmetry operators.

The symmetry operators permissible in molecules are those of *point symmetry*, which leave at least one point of the molecule fixed. The elements of point symmetry are the proper and improper rotation operators. Crystals have, in addition, translational symmetry operators, which generate the periodicity that characterizes the crystalline state. The allowed combinations of point symmetry and translational symmetry operators constitute the *space groups*, of which there are 230.

A general symmetry operation in a crystal may be written

$$\mathbf{y} = \mathbf{R}\mathbf{x} + \mathbf{t} \tag{4–5}$$

where $\mathbf{R}$ is the operator for a proper or improper rotation, and $\mathbf{t}$ is a translation vector. A concise notation for the operator of Eq. (4–5) is

$$\mathbf{y} = \{\mathbf{R}|\mathbf{t}\}\mathbf{x} \tag{4–6}$$

This notation and the properties of the operators were developed by Seitz (1935). The operator $\{\mathbf{I}|\mathbf{t}\}$ symbolizes a pure translation, and $\{\mathbf{R}|\mathbf{0}\}$, where $\mathbf{0}$ is the null vector with zero components, is a pure rotation. When $\mathbf{t}$ is a lattice vector, the infinite set of elements $\{\mathbf{I}|\mathbf{t}\}$ constitutes the crystal's translational group. Similarly, a point group consists of a set of symmetry elements $\{\mathbf{R}|\mathbf{0}\}$ that possesses the group properties mandated above. All the elements $\{\mathbf{R}|\mathbf{0}\}$ are not necessarily elements of the space group, however; for example, there are no twofold rotation axes in space group $P2_1/c$.

There are an infinite number of point groups. The constraints imposed on molecular structure by the requirements of bonding or by steric interactions limit the combinations of symmetry elements actually observed, but molecules can and sometimes do have fivefold rotation axes, for example, or eightfold symmetry, or almost anything else. Except for linear cases, however, the order (the number of elements) of a molecular point group must be finite, and this restriction eliminates such combinations of elements as a fourfold rotation axis perpendicular to a threefold axis (this particular example will be dealt with in Exercise 4–50).

When point group symmetry is coupled with the translational symmetry of a crystal, stringent limitations are placed upon the allowed combinations. For one thing, only one-, two-, three-, four-, and sixfold rotational symmetry is possible in a crystal. Basically, the restriction arises because the rotation operator produces new lattice vectors, which must be compatible with the existing lattice. A simple geometric proof of this restriction is given in *Introduction to Crystallography*; a tensor proof is the subject of Exercise 4–15. A consequence of the limitation of proper or improper rotations to $2\pi/n$ with $n = 1, 2, 3, 4$, or 6 is that there are only 32 point groups that are permissible in crystals.

The goal here is to examine the tensor character of these symmetry operators. Except for a few illustrative examples (see Section 4–8), we shall not attempt derivation or even delineation of the space groups; for the former, see Zachariasen (1945); for the latter, see Volume I of the *International Tables for X-Ray Crystallography* or *Introduction to Crystallography*. Readers interested in representations of space groups are referred to the books of Koster (1957) and Cornwell (1969).

4–2 Combinations of Space Group Operators

Suppose that $\{\mathbf{R}|\mathbf{t}\}$ and $\{\mathbf{S}|\mathbf{u}\}$ are two operators in a space group. Applying $\{\mathbf{R}|\mathbf{t}\}$ to a vector $\mathbf{x}$ generates the vector $\mathbf{y}$ of Eq. (4–5). Applying $\{\mathbf{S}|\mathbf{u}\}$ now to $\mathbf{y}$,

$$\{\mathbf{S}|\mathbf{u}\}\mathbf{y} = \mathbf{S}\mathbf{y} + \mathbf{u} \tag{4–7}$$

Using Eq. (4–5),

$$\{\mathbf{S}|\mathbf{u}\}\{\mathbf{R}|\mathbf{t}\}\mathbf{x} = \mathbf{S}(\mathbf{R}\mathbf{x} + \mathbf{t}) + \mathbf{u} \tag{4–8}$$

$$\{\mathbf{S}|\mathbf{u}\}\{\mathbf{R}|\mathbf{t}\}\mathbf{x} = \{\mathbf{S}\mathbf{R}|\mathbf{S}\mathbf{t} + \mathbf{u}\}\mathbf{x} \tag{4–9}$$

The validity of Eq. (4–9) for any $\mathbf{x}$ is ensured if

$$\{\mathbf{S}|\mathbf{u}\}\{\mathbf{R}|\mathbf{t}\} = \{\mathbf{S}\mathbf{R}|\mathbf{S}\mathbf{t} + \mathbf{u}\} \tag{4–10}$$

An inverse space group operator $\{\mathbf{R}|\mathbf{t}\}^{-1}$ may be defined such that

$$\{\mathbf{R}|\mathbf{t}\}^{-1}\{\mathbf{R}|\mathbf{t}\} = \{\mathbf{I}|\mathbf{0}\} \tag{4–11}$$

where $\{\mathbf{I}|\mathbf{0}\}$ is the space group identity operator. Hence,

$$\{\mathbf{R}|\mathbf{t}\}^{-1} = \{\mathbf{R}^{-1}|-\mathbf{R}^{-1}\mathbf{t}\} \tag{4–12}$$

EXERCISE 4–1 Under what conditions on **R**, **S**, **t**, and **u** will the operators $\{\mathbf{R}|\mathbf{t}\}$ and $\{\mathbf{S}|\mathbf{u}\}$ commute? (Operators **A** and **B** are said to commute if **AB** = **BA**.)

EXERCISE 4–2 Evaluate $\{\mathbf{R}|\mathbf{t}\}\{\mathbf{R}^{-1}| -\mathbf{R}^{-1}\mathbf{t}\}$, thus confirming that Eq. (4–12) does indeed give the inverse operator.

EXERCISE 4–3 Show that $\{\mathbf{R}|\mathbf{t}\}\{\mathbf{I}|\mathbf{t}'\}\{\mathbf{R}|\mathbf{t}\}^{-1} = \{\mathbf{I}|\mathbf{R}\mathbf{t}'\}$. (This result shows that, if $\mathbf{t}'$ is a lattice vector, so is $\mathbf{R}\mathbf{t}'$; see Cornwell, 1969, p. 14.)

EXERCISE 4–4 The symmetry elements of space group $P3_1$ may be generated by repeated application of $\{\mathbf{R}|\mathbf{t}\}$, where $\mathbf{R}$ and $\mathbf{t}$ have the properties $(\mathbf{R})^3 = \mathbf{I}$, $\mathbf{t} = [0, 0, 1/3]$, $\mathbf{R}\mathbf{t} = \mathbf{t}$. Calculate $\{\mathbf{R}|\mathbf{t}\}^3$.

4–3 Transformation of Point Group Operators

Let $\mathbf{y}$ be the vector resulting from the application of the proper or improper rotation operator $\mathbf{R}$ to vector $\mathbf{x}$; that is,

$$\mathbf{y} = \mathbf{R}\mathbf{x} \tag{4–13}$$

In a new coordinate system

$$\mathbf{y}' = \mathbf{R}'\mathbf{x}' \tag{4–14}$$

The coordinate transformation of Eq. (3–12) leads to

$$\mathbf{F}\mathbf{y} = \mathbf{R}'\mathbf{F}\mathbf{x} \tag{4–15}$$

from which

$$\mathbf{y} = \mathbf{F}^{-1}\mathbf{R}'\mathbf{F}\mathbf{x} \tag{4–16}$$

In order that Eqs. (4–13) and (4–16) hold for arbitrary $\mathbf{x}$,

$$\mathbf{R} = \mathbf{F}^{-1}\mathbf{R}'\mathbf{F} \tag{4–17}$$

$$\mathbf{R}' = \mathbf{F}\mathbf{R}\mathbf{F}^{-1} = \mathbf{F}\mathbf{R}\overline{\mathbf{G}} \tag{4–18}$$

$$R'^{i}{}_{j} = F^{i}{}_{k}\, R^{k}{}_{l}\, G_{j}{}^{l} \tag{4–19}$$

This transformation law demonstrates that $\mathbf{R}$ is a second-rank mixed tensor.

EXERCISE 4–5 Show that the trace (sum of the diagonal elements) of the matrix representing a point symmetry operator is invariant under the transformation of Eq. (4–19).

4-4 Transformation of Space Group Operators

The generalized transformation law for a space group operator $\{\mathbf{R}|\mathbf{t}\}$ must accommodate both the transformation of the point symmetry operator $\mathbf{R}$ and a shift of origin. The entire effect of shifting the origin to the point at the end of vector $\mathbf{v}$ is to convert $\{\mathbf{R}|\mathbf{t}\}\mathbf{x}$ to

$$\mathbf{y} = \mathbf{R}\mathbf{x} + \mathbf{t} - \mathbf{v} \tag{4-20}$$

which may be written

$$\mathbf{y} = \mathbf{R}(\mathbf{x} - \mathbf{v}) + \mathbf{R}\mathbf{v} + \mathbf{t} - \mathbf{v} \tag{4-21}$$

The components of vector $\mathbf{x}'$ (in a new coordinate system) are related to those of $\mathbf{x} - \mathbf{v}$ by the usual transformation

$$\mathbf{x}' = \mathbf{F}(\mathbf{x} - \mathbf{v}) \tag{4-22}$$

or, by inversion,

$$\mathbf{x} - \mathbf{v} = \overline{\mathbf{G}}\mathbf{x}' \tag{4-23}$$

Inserting Eq. (4-23) into Eq. (4-21) and multiplying by $\mathbf{F}$,

$$\mathbf{y}' = \mathbf{F}\mathbf{R}\overline{\mathbf{G}}\mathbf{x}' + \mathbf{F}\mathbf{R}\mathbf{v} + \mathbf{F}\mathbf{t} - \mathbf{F}\mathbf{v} \tag{4-24}$$

Therefore, the operator $\{\mathbf{R}'|\mathbf{t}'\}$ is given by

$$\{\mathbf{R}'|\mathbf{t}'\} = \{\mathbf{F}\mathbf{R}\overline{\mathbf{G}}|\mathbf{F}\mathbf{R}\mathbf{v} + \mathbf{F}\mathbf{t} - \mathbf{F}\mathbf{v}\} \tag{4-25}$$

Comparison of Eq. (4-25) with the product

$$\{\mathbf{S}|\mathbf{u}\}\{\mathbf{R}|\mathbf{t}\}\{\mathbf{Q}|\mathbf{s}\} = \{\mathbf{S}\mathbf{R}\mathbf{Q}|\mathbf{S}\mathbf{R}\mathbf{s} + \mathbf{S}\mathbf{t} + \mathbf{u}\} \tag{4-26}$$

leads to the generalized space group transformation law (Seitz, 1935):

$$\{\mathbf{R}'|\mathbf{t}'\} = \{\mathbf{F}| -\mathbf{F}\mathbf{v}\}\{\mathbf{R}|\mathbf{t}\}\{\overline{\mathbf{G}}|\mathbf{v}\} \tag{4-27}$$

EXERCISE 4-6 Evaluate Eq. (4-25) for the special case $\mathbf{R} = \mathbf{I}$.

EXERCISE 4–7 Evaluate Eq. (4–25) for the special case $\mathbf{v} = 0$, $\mathbf{t} = 0$, and compare with Eq. (4–18).

EXERCISE 4–8 Evaluate Eq. (4–25) for the special case of a pure shift of origin, for which $\mathbf{F} = \overline{\mathbf{G}} = \mathbf{I}$.

4–5 Rotations in Cartesian Systems

A right-handed cartesian coordinate system is defined as shown in Fig. 4–1; basis vector $\mathbf{e}_3$ is directed upward normal to the plane of the paper. A positive rotation about $\mathbf{e}_3$ is in the direction that would turn $\mathbf{e}_1$ into $\mathbf{e}_2$—that is, counterclockwise in Fig. 4–1. The components of $\mathbf{u}$ are

$$\mathbf{u} = [w \cos \psi,\ w \sin \psi,\ u^3] \tag{4–28}$$

where w is the length of $\mathbf{u} - [0, 0, u^3]$. Vector $\mathbf{v}$ is produced by rotating vector $\mathbf{u}$ counterclockwise through angle θ; therefore,

$$\mathbf{v} = [w \cos(\psi + \theta),\ w \sin(\psi + \theta),\ u^3] \tag{4–29}$$

Expansion of the trigonometric terms and insertion of the components of $\mathbf{u}$ from Eq. (4–28) yields

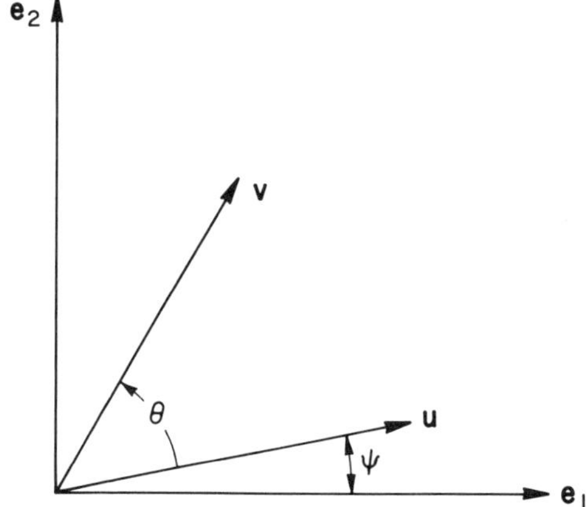

FIGURE 4–1. Direction of positive rotation in a right-handed coordinate system. The direction of $\mathbf{e}_3$ is upward normal to the paper. Rotation of $\mathbf{u}$ counterclockwise through angle θ about $\mathbf{e}_3$ turns it into $\mathbf{v}$.

$$\mathbf{v} = [u^1 \cos\theta - u^2 \sin\theta,\ u^2 \cos\theta + u^1 \sin\theta,\ u^3] \tag{4-30}$$

This is equivalent to the matrix equation

$$\begin{pmatrix} v^1 \\ v^2 \\ v^3 \end{pmatrix} = \begin{pmatrix} \cos\theta & -\sin\theta & 0 \\ \sin\theta & \cos\theta & 0 \\ 0 & 0 & 1 \end{pmatrix} \begin{pmatrix} u^1 \\ u^2 \\ u^3 \end{pmatrix} \tag{4-31}$$

The matrix in Eq. (4–31) represents the operator that rotates a vector counterclockwise through angle θ about $\mathbf{e}_3$; that is,

$$\mathbf{R}_z(\theta) = \begin{pmatrix} \cos\theta & -\sin\theta & 0 \\ \sin\theta & \cos\theta & 0 \\ 0 & 0 & 1 \end{pmatrix} \tag{4-32}$$

Similarly, the matrix operator for a rotation about $\mathbf{e}_1$ in the direction that would move $\mathbf{e}_2$ toward $\mathbf{e}_3$ is

$$\mathbf{R}_x(\theta) = \begin{pmatrix} 1 & 0 & 0 \\ 0 & \cos\theta & -\sin\theta \\ 0 & \sin\theta & \cos\theta \end{pmatrix} \tag{4-33}$$

and the matrix operator for a rotation about $\mathbf{e}_2$ in the direction that would move $\mathbf{e}_3$ toward $\mathbf{e}_1$ is

$$\mathbf{R}_y(\theta) = \begin{pmatrix} \cos\theta & 0 & \sin\theta \\ 0 & 1 & 0 \\ -\sin\theta & 0 & \cos\theta \end{pmatrix} \tag{4-34}$$

4-6 Improper Rotations

In the Schoenflies system of point group symmetry notation, an improper rotation operator $\mathbf{S}_n$ carries out a rotation of $2\pi/n$ coupled with reflection in a plane normal to the rotation axis; the operation consists of *both* the rotation and the reflection. The matrices corresponding to improper rotations about the $\mathbf{e}_3$, $\mathbf{e}_1$, and $\mathbf{e}_2$ axes respectively, of a cartesian coordinate system are obtained by replacing the integer 1 in the matrices for $\mathbf{R}_z(\theta)$, $\mathbf{R}_x(\theta)$, and $\mathbf{R}_y(\theta)$ [Eqs. (4–32), (4–33), (4–34)] by -1.

In the Hermann–Mauguin or International System of notation, the operation that is called an improper rotation is denoted by $\bar{\mathbf{n}}$ and consists of rotation by $2\pi/n$ followed by inversion; the matrices representing these operators are obtained from Eq. (4–32), (4–33), or (4–34) by changing the signs of all the elements.

EXERCISE 4–9 Write the matrices for the following operations in a cartesian coordinate system.

(a) A 90° rotation about $\mathbf{e}_3$.

(b) A 90° rotation about $\mathbf{e}_3$ followed by a 90° rotation about $\mathbf{e}_2$.

(c) A 90° rotation about $\mathbf{e}_2$ followed by a 90° rotation about $\mathbf{e}_3$.

(d) A 120° rotation about $\mathbf{e}_3$.

(e) An $\mathbf{S}_n$ improper rotation about $\mathbf{e}_1$ with $n = 1$.

EXERCISE 4–10 Exercise 4–9(e) showed that $\mathbf{S}_1 = \sigma$, where σ is the Schoenflies symbol for a reflection operator. What operator $\bar{\mathbf{n}}$ corresponds to reflection (symbol m in the Hermann–Mauguin notation)?

EXERCISE 4–11 Write the matrices for all powers of $\mathbf{R}_z(120°)$.

EXERCISE 4–12 (a) Calculate the product matrix of $\mathbf{R}_x(90°)\mathbf{R}_z(120°)$.

(b) Calculate the second, third, and fourth powers of the product $\mathbf{R}_x(90°)\mathbf{R}_z(120°)$.

EXERCISE 4–13 Write expressions in terms of θ for the traces of the matrices that represent proper and improper rotations.

EXERCISE 4–14 Prove that all the elements of the matrix representing a symmetry operation in a crystal are integers if the basis vectors are lattice vectors of a primitive unit cell.

EXERCISE 4–15 Use the results of Exercises 4–13 and 4–14 and the invariance of the trace that was demonstrated in Exercise 4–5 to derive the allowed proper and improper point symmetry operators that can exist in crystals.

4–7 Point Symmetry Operators in General Rectilinear Systems

The definitions and relationships developed in the preceding sections provide the apparatus needed for constructing symmetry operators in the general rectilinear coordinate systems of crystals. Consider, for example, a rotation of 120° about the $\mathbf{a}_3$ axis of a hexagonal crystal. A set of cartesian basis vectors with $\mathbf{e}_3$ parallel to $\mathbf{a}_3$ may be selected by

$$\begin{pmatrix} \mathbf{e}_1 \\ \mathbf{e}_2 \\ \mathbf{e}_3 \end{pmatrix} = \begin{pmatrix} \dfrac{1}{a_1} & 0 & 0 \\ \dfrac{1}{\sqrt{3}a_1} & \dfrac{2}{\sqrt{3}a_1} & 0 \\ 0 & 0 & \dfrac{1}{a_3} \end{pmatrix} \begin{pmatrix} \mathbf{a}_1 \\ \mathbf{a}_2 \\ \mathbf{a}_3 \end{pmatrix} \tag{4–35}$$

By Eq. (4–18), the symmetry operator $\mathbf{R}'$ in the crystal system is

$$\mathbf{R}' = \mathbf{C}_3 = \mathbf{3} = \begin{pmatrix} 0 & -1 & 0 \\ 1 & -1 & 0 \\ 0 & 0 & 1 \end{pmatrix} \tag{4–36}$$

We shall now undertake the determination of the matrix corresponding to a rotation about an arbitrary unit vector $\mathbf{u}$ in a crystal. Having specified $\mathbf{u}$, let us choose unit vectors $\mathbf{s}$ and $\mathbf{t}$ such that $\mathbf{s}$, $\mathbf{t}$, and $\mathbf{u}$ form an orthonormal set. We may transform from the crystal axes to cartesian axes by

$$\begin{pmatrix} \mathbf{e}_1 \\ \mathbf{e}_2 \\ \mathbf{e}_3 \end{pmatrix} = \begin{pmatrix} s^1 & s^2 & s^3 \\ t^1 & t^2 & t^3 \\ u^1 & u^2 & u^3 \end{pmatrix} \begin{pmatrix} \mathbf{a}_1 \\ \mathbf{a}_2 \\ \mathbf{a}_3 \end{pmatrix} \tag{4–37}$$

In briefer notation,

$$\mathbf{e} = \mathbf{G}\mathbf{a} \tag{4–38}$$

$$\mathbf{a} = \mathbf{G}^{-1}\mathbf{e} \tag{4–39}$$

where $\mathbf{G}$ is the matrix in Eq. (4–37), and $\mathbf{G}^{-1}$ is given by

$$\mathbf{G}^{-1} = \begin{pmatrix} s_1 & t_1 & u_1 \\ s_2 & t_2 & u_2 \\ s_3 & t_3 & u_3 \end{pmatrix} \tag{4–40}$$

in which the columns consist of the covariant components of the unit vectors $\mathbf{s}$, $\mathbf{t}$, and $\mathbf{u}$.

A rotation matrix in the crystal system is expressed by

$$\mathbf{R} = \overline{\mathbf{G}}\mathbf{R}'\mathbf{F} \tag{4–41}$$

where as usual $\mathbf{F}$ is contragredient to $\mathbf{G}$ (that is, $\mathbf{F} = \overline{\mathbf{G}}^{-1}$) and $\mathbf{R}'$ is the cartesian rotation matrix $\mathbf{R}_z(\theta)$ of Eq. (4–32). Evaluation of the components of $\mathbf{R}$ leads to (see Exercise 4–16)

$$R^i{}_j = u^i u_j + (\delta^i{}_j - u^i u_j) \cos\theta + g^{ik}\epsilon_{klj}u^l \sin\theta \tag{4–42}$$

The matrix elements for an improper rotation axis (defined as a rotatory-inversion axis according to the Hermann–Mauguin system) are obtained by changing the sign of the right-hand side of Eq. (4–42). The generalized symmetry operator thus has elements

$$R^i{}_j = \pm[u^i u_j + (\delta^i{}_j - u^i u_j)\cos\theta + g^{ik}\epsilon_{klj}u^l \sin\theta] \qquad (4\text{–}43)$$

with the positive sign applying to proper rotations and the negative sign to improper rotations. An improper rotation in the Schoenflies sense (corresponding to an axis of rotatory reflection) may be represented by applying a reflection [taking $\theta = 180°$ and the negative sign in Eq. (4–43)] to the pure rotation of Eq. (4–42).

Dyadic operators were described in Section 3–17. The generalized symmetry operator in dyadic notation (see Zachariasen, 1945; Patterson, 1959) is

$$\mathbf{\Phi} = \pm[\mathbf{uu} + (\mathbf{I} - \mathbf{uu})\cos\theta + \mathbf{I}\,\Lambda\,\mathbf{u}\sin\theta] \qquad (4\text{–}44)$$

In Patterson's formula the last term is $-\mathbf{I}\,\Lambda\,\mathbf{u}\sin\theta$. The nonion form of this dyadic is

$$\mathbf{\Phi} = \pm[u_i u_j + (g_{ij} - u_i u_j)\cos\theta + \epsilon_{ilj}u^l \sin\theta]\mathbf{a}^i\mathbf{a}^j \qquad (4\text{–}45)$$

The quantity in brackets in Eq. (4–45) is of course the doubly covariant component associated with the mixed tensor of Eq. (4–43).

To illustrate the use of these equations, consider a rotation about the $\mathbf{a}_1$ axis of a hexagonal lattice. In this case,

$$\bar{\mathbf{u}} = [1/a_1, 0, 0] = (a_1, -a_1/2, 0) \qquad (4\text{–}46)$$

and Eq. (4–43) gives

$$\mathbf{R} = \begin{pmatrix} 1 & -(1 - \cos\theta)/2 & -a_3\sin\theta/(a_1\sqrt{3}) \\ 0 & \cos\theta & -2a_3\sin\theta/(a_1\sqrt{3}) \\ 0 & a_1\sqrt{3}\sin\theta/(2a_3) & \cos\theta \end{pmatrix} \qquad (4\text{–}47)$$

A symmetry operation in a crystal must generate lattice vectors from lattice vectors. Hence the elements of a symmetry matrix must be rational numbers. Hexagonal symmetry imposes no constraints on the ratio of a_3 to a_1, so the only way that terms such as $a_3\sin\theta/(a_1\sqrt{3})$ in Eq. (4–47) can be restricted to rational numbers is for them to vanish. This shows that the only rotations about [100] compatible with the periodicity of a hexagonal crystal are onefold and twofold. From Eq. (4–47) the twofold rotation matrix is

$$\mathbf{R}(180°) = \begin{pmatrix} 1 & -1 & 0 \\ 0 & -1 & 0 \\ 0 & 0 & -1 \end{pmatrix} \tag{4-48}$$

Table 4-1 lists the matrix representations of the point symmetry operators **1**, **2**, **3**, **4**, and **6** for some of the common orientations referred to the conventional choices of crystal axes.

TABLE 4-1 STANDARD SYMMETRY OPERATORS IN CRYSTAL COORDINATES

Operator	Orientation	Crystal systems	Matrix in direct space	Matrix in reciprocal space
1	Any	Any	$\begin{pmatrix} 1 & 0 & 0 \\ 0 & 1 & 0 \\ 0 & 0 & 1 \end{pmatrix}$	$\begin{pmatrix} 1 & 0 & 0 \\ 0 & 1 & 0 \\ 0 & 0 & 1 \end{pmatrix}$
$\bar{\mathbf{1}}$	Any	Any	$\begin{pmatrix} \bar{1} & 0 & 0 \\ 0 & \bar{1} & 0 \\ 0 & 0 & \bar{1} \end{pmatrix}$	$\begin{pmatrix} \bar{1} & 0 & 0 \\ 0 & \bar{1} & 0 \\ 0 & 0 & \bar{1} \end{pmatrix}$
2	[100]	Orthorhombic Tetragonal Cubic	$\begin{pmatrix} 1 & 0 & 0 \\ 0 & \bar{1} & 0 \\ 0 & 0 & \bar{1} \end{pmatrix}$	$\begin{pmatrix} 1 & 0 & 0 \\ 0 & \bar{1} & 0 \\ 0 & 0 & \bar{1} \end{pmatrix}$
2	[010]	Monoclinic Orthorhombic Tetragonal Cubic	$\begin{pmatrix} \bar{1} & 0 & 0 \\ 0 & 1 & 0 \\ 0 & 0 & \bar{1} \end{pmatrix}$	$\begin{pmatrix} \bar{1} & 0 & 0 \\ 0 & 1 & 0 \\ 0 & 0 & \bar{1} \end{pmatrix}$
2	[100]	Hexagonal Trigonal	$\begin{pmatrix} 1 & \bar{1} & 0 \\ 0 & \bar{1} & 0 \\ 0 & 0 & \bar{1} \end{pmatrix}$	$\begin{pmatrix} 1 & 0 & 0 \\ \bar{1} & \bar{1} & 0 \\ 0 & 0 & \bar{1} \end{pmatrix}$
2	[010]	Hexagonal Trigonal	$\begin{pmatrix} \bar{1} & 0 & 0 \\ \bar{1} & 1 & 0 \\ 0 & 0 & \bar{1} \end{pmatrix}$	$\begin{pmatrix} \bar{1} & \bar{1} & 0 \\ 0 & 1 & 0 \\ 0 & 0 & \bar{1} \end{pmatrix}$
2	[110]	Hexagonal Trigonal	$\begin{pmatrix} 0 & 1 & 0 \\ 1 & 0 & 0 \\ 0 & 0 & \bar{1} \end{pmatrix}$	$\begin{pmatrix} 0 & 1 & 0 \\ 1 & 0 & 0 \\ 0 & 0 & \bar{1} \end{pmatrix}$
2	[210]	Hexagonal Trigonal	$\begin{pmatrix} 1 & 0 & 0 \\ 1 & \bar{1} & 0 \\ 0 & 0 & \bar{1} \end{pmatrix}$	$\begin{pmatrix} 1 & 1 & 0 \\ 0 & \bar{1} & 0 \\ 0 & 0 & \bar{1} \end{pmatrix}$
2	[1$\bar{1}$0]	Hexagonal Trigonal	$\begin{pmatrix} 0 & \bar{1} & 0 \\ \bar{1} & 0 & 0 \\ 0 & 0 & \bar{1} \end{pmatrix}$	$\begin{pmatrix} 0 & \bar{1} & 0 \\ \bar{1} & 0 & 0 \\ 0 & 0 & \bar{1} \end{pmatrix}$
2	[120]	Hexagonal Trigonal	$\begin{pmatrix} \bar{1} & 1 & 0 \\ 0 & 1 & 0 \\ 0 & 0 & \bar{1} \end{pmatrix}$	$\begin{pmatrix} \bar{1} & 0 & 0 \\ 1 & 1 & 0 \\ 0 & 0 & \bar{1} \end{pmatrix}$

TABLE 4-1 (*Continued)*

Operator	Orientation	Crystal systems	Matrix in direct space	Matrix in reciprocal space
2	[110]	Tetragonal Cubic	$\begin{pmatrix} 0 & 1 & 0 \\ 1 & 0 & 0 \\ 0 & 0 & \bar{1} \end{pmatrix}$	$\begin{pmatrix} 0 & 1 & 0 \\ 1 & 0 & 0 \\ 0 & 0 & \bar{1} \end{pmatrix}$
2	$[1\bar{1}0]$	Tetragonal Cubic	$\begin{pmatrix} 0 & \bar{1} & 0 \\ \bar{1} & 0 & 0 \\ 0 & 0 & \bar{1} \end{pmatrix}$	$\begin{pmatrix} 0 & \bar{1} & 0 \\ \bar{1} & 0 & 0 \\ 0 & 0 & \bar{1} \end{pmatrix}$
2	$[1\bar{1}0]$	Rhombohedral	$\begin{pmatrix} 0 & \bar{1} & 0 \\ \bar{1} & 0 & 0 \\ 0 & 0 & \bar{1} \end{pmatrix}$	$\begin{pmatrix} 0 & \bar{1} & 0 \\ \bar{1} & 0 & 0 \\ 0 & 0 & \bar{1} \end{pmatrix}$
2	$[10\bar{1}]$	Rhombohedral	$\begin{pmatrix} 0 & 0 & \bar{1} \\ 0 & \bar{1} & 0 \\ \bar{1} & 0 & 0 \end{pmatrix}$	$\begin{pmatrix} 0 & 0 & \bar{1} \\ 0 & \bar{1} & 0 \\ \bar{1} & 0 & 0 \end{pmatrix}$
2	$[01\bar{1}]$	Rhombohedral	$\begin{pmatrix} \bar{1} & 0 & 0 \\ 0 & 0 & \bar{1} \\ 0 & \bar{1} & 0 \end{pmatrix}$	$\begin{pmatrix} \bar{1} & 0 & 0 \\ 0 & 0 & \bar{1} \\ 0 & \bar{1} & 0 \end{pmatrix}$
2	[001]	Orthorhombic Hexagonal Tetragonal Cubic	$\begin{pmatrix} \bar{1} & 0 & 0 \\ 0 & \bar{1} & 0 \\ 0 & 0 & 1 \end{pmatrix}$	$\begin{pmatrix} \bar{1} & 0 & 0 \\ 0 & \bar{1} & 0 \\ 0 & 0 & 1 \end{pmatrix}$
3	[001]	Hexagonal Trigonal	$\begin{pmatrix} 0 & \bar{1} & 0 \\ 1 & \bar{1} & 0 \\ 0 & 0 & 1 \end{pmatrix}$	$\begin{pmatrix} \bar{1} & \bar{1} & 0 \\ 1 & 0 & 0 \\ 0 & 0 & 1 \end{pmatrix}$
$\mathbf{3}^2$	[001]	Hexagonal Trigonal	$\begin{pmatrix} \bar{1} & 1 & 0 \\ \bar{1} & 0 & 0 \\ 0 & 0 & 1 \end{pmatrix}$	$\begin{pmatrix} 0 & 1 & 0 \\ \bar{1} & \bar{1} & 0 \\ 0 & 0 & 1 \end{pmatrix}$
3	[111]	Rhombohedral Cubic	$\begin{pmatrix} 0 & 0 & 1 \\ 1 & 0 & 0 \\ 0 & 1 & 0 \end{pmatrix}$	$\begin{pmatrix} 0 & 0 & 1 \\ 1 & 0 & 0 \\ 0 & 1 & 0 \end{pmatrix}$
$\mathbf{3}^2$	[111]	Rhombohedral Cubic	$\begin{pmatrix} 0 & 1 & 0 \\ 0 & 0 & 1 \\ 1 & 0 & 0 \end{pmatrix}$	$\begin{pmatrix} 0 & 1 & 0 \\ 0 & 0 & 1 \\ 1 & 0 & 0 \end{pmatrix}$
4	[001]	Tetragonal Cubic	$\begin{pmatrix} 0 & \bar{1} & 0 \\ 1 & 0 & 0 \\ 0 & 0 & 1 \end{pmatrix}$	$\begin{pmatrix} 0 & \bar{1} & 0 \\ 1 & 0 & 0 \\ 0 & 0 & 1 \end{pmatrix}$
$\mathbf{4}^3$	[001]	Tetragonal Cubic	$\begin{pmatrix} 0 & 1 & 0 \\ \bar{1} & 0 & 0 \\ 0 & 0 & 1 \end{pmatrix}$	$\begin{pmatrix} 0 & 1 & 0 \\ \bar{1} & 0 & 0 \\ 0 & 0 & 1 \end{pmatrix}$

TABLE 4-1 *(Continued)*

Operator	Orientation	Crystal systems	Matrix in direct space	Matrix in reciprocal space
6	[001]	Hexagonal Trigonal	$\begin{pmatrix} 1 & \bar{1} & 0 \\ 1 & 0 & 0 \\ 0 & 0 & 1 \end{pmatrix}$	$\begin{pmatrix} 0 & \bar{1} & 0 \\ 1 & 1 & 0 \\ 0 & 0 & 1 \end{pmatrix}$
$\mathbf{6}^5$	[001]	Hexagonal Trigonal	$\begin{pmatrix} 0 & 1 & 0 \\ \bar{1} & 1 & 0 \\ 0 & 0 & 1 \end{pmatrix}$	$\begin{pmatrix} 1 & 1 & 0 \\ \bar{1} & 0 & 0 \\ 0 & 0 & 1 \end{pmatrix}$

EXERCISE 4-16 Carry out all steps of the algebra leading from Eq. (4-41) to the components of Eq. (4-42).

EXERCISE 4-17 Derive Eq. (4-45) from Eq. (4-44). (See Exercise 3-65 for $\mathbf{I} \wedge \mathbf{u}$.)

EXERCISE 4-18 Use Eq. (4-43) to evaluate in a general rectilinear system the matrices for the operators (a) **1**, (b) **2**, (c) **m**, (d) $\bar{\mathbf{1}}$.

EXERCISE 4-19 Derive the matrix for a rotation through angle θ about $\mathbf{a}_2$ of a monoclinic crystal. What restrictions on θ must be imposed if this matrix is to be a symmetry operator of the monoclinic system?

EXERCISE 4-20 Calculate all powers of the **3** operator in a hexagonal crystal [see Eq. (4-36)].

EXERCISE 4-21 Two cartesian systems are related by the transformation

$$\begin{pmatrix} \mathbf{e}_1' \\ \mathbf{e}_2' \\ \mathbf{e}_3' \end{pmatrix} = \begin{pmatrix} 1/\sqrt{2} & -1/\sqrt{2} & 0 \\ 1/\sqrt{6} & 1/\sqrt{6} & -2/\sqrt{6} \\ 1/\sqrt{3} & 1/\sqrt{3} & 1/\sqrt{3} \end{pmatrix} \begin{pmatrix} \mathbf{e}_1 \\ \mathbf{e}_2 \\ \mathbf{e}_3 \end{pmatrix}$$

Apply this transformation to Eq. (4-32) to determine the matrix that represents a 120° rotation about the body diagonal of a cube.

EXERCISE 4-22 Use Eq. (4-42) to derive a matrix for a rotation through angle θ about [111] of a cubic crystal.

EXERCISE 4-23 Determine the matrix representing a mirror plane normal to the $\mathbf{a}_1$ axis of a hexagonal crystal.

EXERCISE 4-24 Apply Eq. (4-42) to find the matrix that represents rotation through angle θ about

(a) [100] in an orthorhombic crystal;
(b) [001] in a tetragonal crystal;

(c) [001] in a hexagonal crystal;
(d) [210] in a hexagonal crystal;
(e) [111] in a rhombohedral crystal;
(f) [110] in a cubic crystal.

Exercise 4–25 For each of the cases treated in Exercise 4–24, what values of θ are consistent with the translational symmetry?

Exercise 4–26 The matrix for a center of inversion at the origin is $-\mathbf{I}$. Show that this matrix is invariant under any transformation.

4–8 Space Group Derivation

The derivation of equivalent positions from Hermann–Mauguin space group symbols is a straightforward process. The process is especially simple when the symmetry elements all pass through the origin, but the methods that have been developed in the preceding sections permit extension to cases where the symmetry elements are neither aligned along the axes nor constrained to pass through the origin.

It will be helpful to establish first the general relationship between points related by a center of inversion. Suppose that for each point x^1, x^2, x^3 there is an equivalent point with coordinates $t^1 - x^1$, $t^2 - x^2$, $t^3 - x^3$. A shift of origin to the point $t^1/2$, $t^2/2$, $t^3/2$ changes the coordinates of the first point to $x^1 - t^1/2$, $x^2 - t^2/2$, $x^3 - t^3/2$, and the coordinates of the point related by symmetry are $t^1/2 - x^1$, $t^2/2 - x^2$, $t^3/2 - x^3$. Thus, there is a center of inversion at the point $t^1/2$, $t^2/2$, $t^3/2$ (see Exercise 4–26).

The techniques of space group derivation will be illustrated by means of examples. Consider the derivation of the equivalent positions of space group $P2_1/n$ with the origin at a center of inversion. If the $\mathbf{2_1}$ screw axis and the $\mathbf{n}$ glide plane go through the origin, the positions are

$$x, y, z; \bar{x}, \tfrac{1}{2} + y, \bar{z}; \tfrac{1}{2} + x, \bar{y}, \tfrac{1}{2} + z; \tfrac{1}{2} - x, \tfrac{1}{2} - y, \tfrac{1}{2} - z$$

in which we follow the standard convention of denoting the coordinates by x, y, z. According to the result derived in the previous paragraph, there is a center of inversion at $\mathbf{v} = [1/4, 1/4, 1/4]$. The origin may be shifted to the center of inversion by setting $\mathbf{F} = \mathbf{G} = \mathbf{I}$ and $\mathbf{v} = [1/4, 1/4, 1/4]$ in Eq. (4–27). The four equivalent positions of space group $P2_1/n$ with the origin at the center of inversion are therefore

$$x, y, z;\ \tfrac{1}{2} - x, \tfrac{1}{2} + y, \tfrac{1}{2} - z;\ \tfrac{1}{2} + x, \tfrac{1}{2} - y, \tfrac{1}{2} + z;\ \bar{x}, \bar{y}, \bar{z}$$

As another example, we shall derive the 16-fold general positions of space group $P4_2/ncm$ for three different choices of origin: (a) origin at the intersection of the **n** glide plane (which is normal to $\mathbf{a}_3$ with glide component [1/2, 1/2, 0]), the **c** glide plane (which is normal to $\mathbf{a}_1$), and the mirror plane (which is normal to [1 1 0]); (b) origin shifted from that in (a) to [1/2, 0, 1/4], which places it at a position with symmetry $\bar{4}$ (this is one of the origin choices listed in the *International Tables for X-Ray Crystallography*); (c) origin at the center of symmetry whose full symmetry is $2/m$ (this is the second origin choice tabulated in *International Tables for X-Ray Crystallography*). The results are tabulated in Table 4-2. In column (a) of Table 4-2, the second entry is the result of applying operator $\mathbf{4}_2$ to x, y, z; that is, the operator is

$$\mathbf{4_2} = \{\mathbf{R}|\mathbf{t}\} = \left\{ \begin{pmatrix} 0 & \bar{1} & 0 \\ 1 & 0 & 0 \\ 0 & 0 & 1 \end{pmatrix} \middle| \begin{pmatrix} 0 \\ 0 \\ \tfrac{1}{2} \end{pmatrix} \right\} \tag{4-49}$$

TABLE 4-2 GENERAL POSITIONS OF SPACE GROUP $P4_2/ncm$

	(a) Origin on $\mathbf{4_2}$ and $\mathbf{m}$	(b) Origin at $\bar{\mathbf{4}}$	(c) Origin at $\mathbf{2/m}$
(1)	x, y, z	x, y, z	x, y, z
(2)	$\bar{y}, x, \tfrac{1}{2} + z$	$\tfrac{1}{2} - y, \tfrac{1}{2} + x, \tfrac{1}{2} + z$	$\tfrac{1}{2} - y, x, \tfrac{1}{2} + z$
(3)	$\bar{x}, \bar{y}, z$	$\bar{x}, \bar{y}, z$	$\tfrac{1}{2} - x, \tfrac{1}{2} - y, z$
(4)	$y, \bar{x}, \tfrac{1}{2} + z$	$\tfrac{1}{2} + y, \tfrac{1}{2} - x, \tfrac{1}{2} + z$	$y, \tfrac{1}{2} - x, \tfrac{1}{2} + z$
(5)	$\tfrac{1}{2} + x, \tfrac{1}{2} + y, \bar{z}$	$\tfrac{1}{2} + x, \tfrac{1}{2} + y, \tfrac{1}{2} - z$	$\tfrac{1}{2} + x, \tfrac{1}{2} + y, \bar{z}$
(6)	$\tfrac{1}{2} - y, \tfrac{1}{2} + x, \tfrac{1}{2} - z$	$\bar{y}, x, \bar{z}$	$\bar{y}, \tfrac{1}{2} + x, \tfrac{1}{2} - z$
(7)	$\tfrac{1}{2} - x, \tfrac{1}{2} - y, \bar{z}$	$\tfrac{1}{2} - x, \tfrac{1}{2} - y, \tfrac{1}{2} - z$	$\bar{x}, \bar{y}, \bar{z}$
(8)	$\tfrac{1}{2} + y, \tfrac{1}{2} - x, \tfrac{1}{2} - z$	$y, \bar{x}, \bar{z}$	$\tfrac{1}{2} + y, \bar{x}, \tfrac{1}{2} - z$
(9)	$\bar{x}, y, \tfrac{1}{2} + z$	$\bar{x}, y, \tfrac{1}{2} + z$	$\tfrac{1}{2} - x, y, \tfrac{1}{2} + z$
(10)	$\bar{y}, \bar{x}, z$	$\tfrac{1}{2} - y, \tfrac{1}{2} - x, z$	$\tfrac{1}{2} - y, \tfrac{1}{2} - x, z$

TABLE 4-2 (*Continued*)

	(a) Origin on $\mathbf{4_2}$ and **m**	(b) Origin at $\bar{\mathbf{4}}$	(c) Origin at **2/m**
(11)	$x, \bar{y}, \frac{1}{2} + z$	$x, \bar{y}, \frac{1}{2} + z$	$x, \frac{1}{2} - y, \frac{1}{2} + z$
(12)	y, x, z	$\frac{1}{2} + y, \frac{1}{2} + x, z$	y, x, z
(13)	$\frac{1}{2} - x, \frac{1}{2} + y, \frac{1}{2} - z$	$\frac{1}{2} - x, \frac{1}{2} + y, \bar{z}$	$\bar{x}, \frac{1}{2} + y, \frac{1}{2} - z$
(14)	$\frac{1}{2} - y, \frac{1}{2} - x, \bar{z}$	$\bar{y}, \bar{x}, \frac{1}{2} - z$	$\bar{y}, \bar{x}, \bar{z}$
(15)	$\frac{1}{2} + x, \frac{1}{2} - y, \frac{1}{2} - z$	$\frac{1}{2} + x, \frac{1}{2} - y, \bar{z}$	$\frac{1}{2} + x, \bar{y}, \frac{1}{2} - z$
(16)	$\frac{1}{2} + y, \frac{1}{2} + x, \bar{z}$	$y, x, \frac{1}{2} - z$	$\frac{1}{2} + y, \frac{1}{2} + x, \bar{z}$

Entries 3 and 4 in column (a) are powers of $\mathbf{4_2}$. Entry 5 is **n**, and entries 6, 7, and 8 are the products of $\mathbf{4_2}$, $(\mathbf{4_2})^2$, and $(\mathbf{4_2})^3$ with **n**. Entries 9–12 are **c**, $\mathbf{4_2c}$, $(\mathbf{4_2})^2\mathbf{c}$, and $(\mathbf{4_2})^3$c. Entry 13 is **nc**, and entries 14–16 are $\mathbf{4_2nc}$, $(\mathbf{4_2})^2\mathbf{nc}$, and $(\mathbf{4_2})^3\mathbf{nc}$. Columns (b) and (c) of Table 4-2 were derived from column (a) by setting $\mathbf{F} = \mathbf{G} = \mathbf{I}$ in Eq. (4–27) or (4–25). For example, entry 4 in column (b), the positions generated by $(\mathbf{4_2})^3$, are derived by putting

$$\mathbf{R} = \begin{pmatrix} 0 & 1 & 0 \\ \bar{1} & 0 & 0 \\ 0 & 0 & 1 \end{pmatrix}, \quad \mathbf{v} = \begin{pmatrix} \frac{1}{2} \\ 0 \\ \frac{1}{4} \end{pmatrix}, \quad \mathbf{t} = \begin{pmatrix} 0 \\ 0 \\ \frac{1}{2} \end{pmatrix} \tag{4–50}$$

into Eq. (4–25). The result of the calculation is $-1/2 + y$, $-1/2 - x$, $1/2 + z$, which, by the modulo 1 arithmetic that is permitted for the periodic contravariant fractional coordinates, is equivalent to $1/2 + y$, $1/2 - x$, $1/2 + z$. Similarly, the entries in column (c) are obtained from those in column (a) by taking $\mathbf{v} = [1/4, 1/4, 0]$, or from those in column (b) by taking $\mathbf{v} = [-1/4, 1/4, -1/4]$.

Exercise 4–27 Verify the entries in row 13 of Table 4–2 for the equivalent positions generated by **nc**.

Exercise 4–28 Derive the 12-fold general positions of space group $P6/m$ (a) with the origin at **6/m**; (b) with the origin shifted to $\bar{\mathbf{6}}$, at [2/3, 1/3, 0] from that of (a).

Exercise 4–29 Derive the 12-fold general positions of space group $P\bar{3}m1$ (a) with the origin at $\bar{\mathbf{3}}\mathbf{m}$; (b) with the origin shifted to the $2/m$ position [1/2, 0, 0].

4-9 Point Symmetry in Reciprocal Space

Equation (4–13) is a linear transformation by means of which one set of contravariant coordinates is converted into another. As was shown in Section 3–3, the transformation matrix for covariant quantities is the contragredient matrix of the contravariant transformation. Thus, if **R** is a symmetry operator for contravariant components, the matrix $\overline{\mathbf{R}}^{-1}$ represents the symmetry operator for covariant components. To illustrate, the matrix operator for a threefold rotation axis about $\mathbf{a}_3$ of a hexagonal crystal was given by Eq. (4–36); one application of this operator converted x^1, x^2, x^3 into $-x^2$, $x^1 - x^2$, x^3, a second application produces $x^2 - x^1$, $-x^1$, x^3, and a third application returns to the original point. In reciprocal space the threefold operator is obtained by taking the transpose of the reciprocal of the matrix in Eq. (4–36). Thus,

$$\mathbf{C}_3{}^* = \overline{\mathbf{C}}_3{}^{-1} = \begin{pmatrix} -1 & -1 & 0 \\ 1 & 0 & 0 \\ 0 & 0 & 1 \end{pmatrix} \tag{4–51}$$

This operator will convert covariant components h_1, h_2, h_3 into covariant components $-h_1 - h_2$, h_1, h_3; a second application will produce h_2, $-h_1 - h_2$, h_3; and a third application will get back to h_1, h_2, h_3. An important consequence of this result is that, when an operator **R** produces sets of equivalent positions at which atoms might be located, the contragredient symmetry operator

$$\mathbf{R}^* = \overline{\mathbf{R}}^{-1} \tag{4–52}$$

produces sets of equivalent reciprocal vectors, which correspond to equivalent Bragg reflections. Thus, the threefold symmetry operator of a hexagonal lattice generates equivalent reflections h_1, h_2, h_3; $-h_1 - h_2$, h_1, h_3; h_2, $-h_1 - h_2$, h_3. [Note: Crystallographers often use a four-index notation (*hkil*) for hexagonal crystals, in which h, k, and l correspond to our h_1, h_2, and h_3, and $i = -h - k$; this permits generation of the indices of equivalent reflections by cyclic permutation of *hki* to give (*hkil*), (*kihl*), (*ihkl*).]

The collection of elements of the symmetry group of the reciprocal space of a crystal may in fact be deduced in an even easier fashion than taking the contragredient of each element of the symmetry group of the direct space. Suppose that **R** is an element of the direct space symmetry group. The group properties of Section 4–1 guarantee that $\mathbf{R}^{-1}$ is also an element of the direct space symmetry group. But the contragredient of each of these is an element of the reciprocal space symmetry group. Thus, $\overline{\mathbf{R}}$ is an element of the reciprocal space symmetry group. If we have a set of matrices $\mathbf{R}(i)$ that constitute the direct space symmetry group, the set of matrices $\overline{\mathbf{R}}(j)$ constitute the reciprocal space symmetry group. This does not require, however, that **R** and $\mathbf{R}^* = \overline{\mathbf{R}}$ pertain to the same symmetry element; in our example of a 120° rotation about the hexagonal $\mathbf{a}_3$, the transpose of **R** gave the operator for a 240° rotation in reciprocal space.

The diffraction pattern of a crystal must have at least the symmetry described by the set of elements $\mathbf{R}^*$. If the radiation is not near to an absorption edge of the scattering atoms, the diffraction pattern is centrosymmetric. The centrosymmetric crystallographic point groups are known as *Laue groups.* The Laue group of a centrosymmetric crystal is the same as its point group, and the full set of matrices $\mathbf{R}^*$ is obtained by transposing the set of matrices **R**. The matrices for the Laue group of a noncentrosymmetric crystal include all the $\overline{\mathbf{R}}$ matrices plus the matrices obtained by changing the sign of each $\overline{\mathbf{R}}$; thus, the order of the Laue group of a noncentrosymmetric crystal is twice the order of the point group.

Exercise 4–30 From the general positions of space group $P6/m$ derived in Exercise 4–28, derive the entire set of equivalent reflections.

Exercise 4–31 Derive the entire set of equivalent reflections for space group $P6/m$ by applying the suffix-raising and suffix-lowering property of the metric tensors, $\mathbf{R}^* = \mathbf{g}\mathbf{R}\mathbf{g}^*$, to the general positions.

Exercise 4–32 Derive the list of equivalent general positions and equivalent reflections for space group $P\bar{4}2m$, which has a $\bar{4}$ axis along the tetragonal $\mathbf{a}_3$, a twofold axis along $\mathbf{a}_1$, and a mirror plane normal to [110].

4–10 Effect of Symmetry on Second-Rank Tensors

A point symmetry operator **R** has, in common with the coordinate transformation matrix **F**, the property of generating new contravariant coordinates from the old. The differences between **R** and **F** are that (1) **R** operates within a fixed coordinate system, whereas **F** expresses a relationship between the coordinates of a fixed point referred to two different sets of basis vectors, and (2) **R** is a tensor, but **F** generally is not. Other than these distinctions, however, **R** and **F** play similar roles, and much of the transformation theory developed in Chapter 3 can be transferred directly to symmetry operators.

A particular extension of transformation theory to symmetry is suggested by the metric tensor transformation of Eq. (3–32). Since, however, the symmetry operation relates points within a single coordinate system, the metric tensor must be unchanged by the operation; hence,

$$\mathbf{g} = \overline{\mathbf{R}}^{-1}\mathbf{g}\mathbf{R}^{-1} = \overline{\mathbf{R}}\mathbf{g}\mathbf{R} \tag{4–53}$$

Equation (4–53) imposes constraints on the unit cell parameters that are consistent with particular symmetry operators.

In general, the tensor representing any physical property is invariant under the operations of the relevant symmetry group. If, for example, β is the doubly contravariant tensor representing the anisotropic temperature factor of an atom in a crystal, the anisotropic temperature factor tensor of a second atom related to the first by point symmetry operator **R** is given by the equation analogous to Eq. (3–104),

$$\beta' = \mathbf{R}\beta\overline{\mathbf{R}} \tag{4–54}$$

If the symmetry operator passes through an atom, the operation produces the same atom; in this case β' is equal to β, and the symmetry-imposed constraints on the components of β are given by the analogue of Eq. (4–53):

$$\beta = \mathbf{R}\beta\overline{\mathbf{R}} \tag{4–55}$$

$$\beta^{ij} = R^i{}_k\, \beta^{kl} R^j{}_l \tag{4–56}$$

The parallel relationships for a doubly covariant tensor **A** are

$$\mathbf{A} = \mathbf{R}^*\mathbf{A}\overline{\mathbf{R}}^* \tag{4–57}$$

$$A_{ij} = R^*{}_i{}^k A_{kl} R^*{}_j{}^l \quad (4\text{–}58)$$

where $\mathbf{R}^*$ is a symmetry operator referred to reciprocal axes. In terms of $\overline{\mathbf{R}}$,

$$\mathbf{A} = \mathbf{gRg^*Ag^*\overline{R}g} \quad (4\text{–}59)$$

EXERCISE 4–33 A symmetry operator is represented by the matrix

$$\mathbf{R} = \begin{pmatrix} -1 & 1 & 0 \\ -1 & 0 & 0 \\ 0 & 0 & 1 \end{pmatrix}$$

Find the restrictions on the components of $\mathbf{g}$ imposed by Eq. (4–53) with this operator.

EXERCISE 4–34 Equation (4–36) gives the matrix for the operator $\mathbf{C}_3$ in a hexagonal crystal.

(a) Find the relationships between the anisotropic temperature factor components β^{ij} of two atoms that are related by the $\mathbf{C}_3$ operator in a hexagonal crystal.

(b) What constraints are imposed by symmetry upon the components of β of an atom that lies on the $\mathbf{C}_3$ axis?

EXERCISE 4–35 Consider a $\mathbf{C}_2$ operator aligned along $\mathbf{a}_1$ in a hexagonal crystal.

(a) Find the relationships between the components β^{ij} of two atoms that are related by this operator.

(b) What constraints are imposed by this operator upon the components of β of an atom that lies on this $\mathbf{C}_2$ axis?

EXERCISE 4–36 (a) Find the relationships that exist between the components of β of two atoms that are related by a $\mathbf{C}_4$ operator in a tetragonal crystal.

(b) What constraints does this symmetry operator impose upon the components of β for an atom that lies on the $\mathbf{C}_4$ axis?

EXERCISE 4–37 (a) Find the relationships that exist between the components of β of two atoms that are related by a $\mathbf{C}_2$ operator that is aligned along $\mathbf{a}_1$ in a tetragonal crystal.

(b) What conditions are imposed upon the components of β of an atom that lies on this $\mathbf{C}_2$ axis?

EXERCISE 4–38 The matrix operator for a reflection plane normal to the [110] direction of a tetragonal crystal is

$$\mathbf{R} = \begin{pmatrix} 0 & \bar{1} & 0 \\ \bar{1} & 0 & 0 \\ 0 & 0 & 1 \end{pmatrix}$$

(a) Find the relationships between the components of β of two atoms that are related by this symmetry operator.

(b) What restrictions exist among the components of β of an atom that lies upon this mirror plane?

EXERCISE 4-39 Find the relationships that exist between the components of β of two atoms that are related by a $\mathbf{C}_3$ operator aligned along [111] in a cubic crystal. See Exercise 4-21 for derivation of the operator. What are the conditions that exist on the components of β of an atom that lies on this $\mathbf{C}_3$ axis?

EXERCISE 4-40 Find the symmetry-imposed conditions that exist among the components of β of an atom that occupies a site of $C_{2h} - 2/m$ symmetry in a monoclinic crystal.

EXERCISE 4-41 An atom occupies a site of $C_{2v} - mm2$ symmetry in an orthorhombic crystal. Find the conditions that this symmetry imposes upon the components of the anisotropic temperature factor β of the atom.

EXERCISE 4-42 What is the relationship analogous to Eq. (4-53) that pertains to $\mathbf{g}^*$?

EXERCISE 4-43 Prove that in any rectilinear coordinate system the determinant R of a point symmetry operator is $+1$ for a proper rotation and -1 for an improper rotation.

4-11 Effect of Symmetry on Vector Products

In Section 3-12 it was shown that vector products transform under change of axes in the same way as ordinary contravariant vectors, provided that consideration is given to the effect on the permutation tensor of any change in the handedness of the basis vectors. In the case of symmetry, the basis vectors and the permutation tensor are both invariant under the operations, and an intrinsic difference appears between the axial vectors generated by the vector product and the ordinary polar vectors. The contravariant components w^m of the vector product $\mathbf{u} \wedge \mathbf{v}$ were given by Eq. (1-80) as $\epsilon_{jkl} u^k v^l g^{jm}$. Application of point symmetry operator $\mathbf{R}$ to $\mathbf{u}$ and $\mathbf{v}$ produces new vectors, say $\mathbf{s}$ and $\mathbf{t}$, from $\mathbf{u}$ and $\mathbf{v}$, respectively. That is,

$$s^i = R^i_{\ j} u^j \tag{4-60}$$

$$t^k = R^k{}_l \, v^l \tag{4–61}$$

The vector product $\mathbf{s} \wedge \mathbf{t}$ will be labeled $\mathbf{r}$, and our goal is to determine the relationship between $\mathbf{w}$ and $\mathbf{r}$.

According to Eq. (1–80), the contravariant components of $\mathbf{r}$ are

$$r^l = \epsilon_{ijk} s^i t^j g^{kl} \tag{4–62}$$

Although neither ϵ nor $\mathbf{g}$ is affected by the symmetry operation, the derivation will be facilitated by replacing g^{kl} by the result of Exercise 4–42—that is, by $R^k{}_m \, R^l{}_n \, g^{mn}$. Then, using Eqs. (4–60) and (4–61),

$$r^l = \epsilon_{ijk} R^i{}_p \, u^p R^j{}_q \, v^q R^k{}_m \, R^l{}_n \, g^{mn} \tag{4–63}$$

By the determinant property of Section 3–11,

$$r^l = R R^l{}_n \, \epsilon_{pqm} u^p v^q g^{mn} \tag{4–64}$$

and by Eq. (1–80)

$$r^l = R R^l{}_n \, w^n \tag{4–65}$$

where R is the determinant of $\mathbf{R}$.

This result reveals a real difference between an axial vector such as $\mathbf{u} \wedge \mathbf{v}$ and polar vectors. The vector product is defined always in terms of a right-handed screw operation. Transformation to a new set of basis vectors leaves $\mathbf{u}$, $\mathbf{v}$, and $\mathbf{u} \wedge \mathbf{v}$ unchanged in direction, although their components must be referred to the new axes. A symmetry operator applies to both $\mathbf{u}$ and $\mathbf{v}$, and the application of an improper rotation to both produces an effective proper rotation, $R\mathbf{R}$.

EXERCISE 4–44 Sketch diagrams showing the effect on $\mathbf{u}$, $\mathbf{v}$, and $\mathbf{u} \wedge \mathbf{v}$ of (a) a $\bar{\mathbf{1}}$ inversion operator, (b) a $\bar{\mathbf{2}} = \mathbf{m}$ operator with $\mathbf{u}$ and $\mathbf{v}$ lying in the mirror plane.

EXERCISE 4–45 Given $\mathbf{u} = [3, 4, -1]$, $\mathbf{v} = [5, -2, 2]$ in a hexagonal crystal:

(a) Calculate the contravariant components of $\mathbf{w} = \mathbf{u} \wedge \mathbf{v}$ in terms of the unit cell dimensions.

(b) Apply the symmetry operator $\bar{\mathbf{3}}$ given by the matrix

$$\begin{pmatrix} 0 & 1 & 0 \\ -1 & 1 & 0 \\ 0 & 0 & -1 \end{pmatrix} \quad \text{to } \mathbf{u} \text{ and } \mathbf{v} \text{ to obtain } \mathbf{s} = \mathbf{Ru}, \mathbf{t} = \mathbf{Rv}.$$

(c) Calculate the components of $\mathbf{r} = \mathbf{s} \wedge \mathbf{t}$ and compare with the value predicted by Eq. (4–65).

4–12 Principal Axes of Symmetry Operators

Section 3–14 developed techniques for finding the principal axes or eigenvectors of second-rank tensors in rectilinear coordinate systems. In all such calculations it is crucial that the covariant and contravariant properties of the tensors be observed carefully. For symmetry operator **R** the eigenvalue equation [Eq. (3–114)] becomes

$$\mathbf{Rv} = \lambda \mathbf{v} \tag{4–66}$$

In suffix notation

$$R^j{}_k \, v^k = \lambda v^j \tag{4–67}$$

and the problem is simplified by already having contravariant vector components on both sides of the equation. The secular equation is

$$|\mathbf{R} - \lambda \mathbf{I}| = 0 \tag{4–68}$$

or

$$|R^j{}_k - \lambda \delta^j{}_k| = 0 \tag{4–69}$$

To illustrate the method, consider the rotation about $\mathbf{e}_3$ given by the cartesian operator of Eq. (4–32). The secular equation is

$$\begin{vmatrix} \cos\theta - \lambda & -\sin\theta & 0 \\ \sin\theta & \cos\theta - \lambda & 0 \\ 0 & 0 & 1 - \lambda \end{vmatrix} = 0 \tag{4–70}$$

One root of this is $\lambda = 1$, and the associated eigenvector is [0, 0, 1]. The other two eigenvalues are $e^{i\theta}$ and $e^{-i\theta}$, and their respective eigenvectors if $\sin\theta \neq 0$ are $[1/\sqrt{2}, -i/\sqrt{2}, 0]$ and $[1/\sqrt{2}, i/\sqrt{2}, 0]$, The orthonormality condition for complex vectors **u** and **v** is

$$\mathbf{u}^* \cdot \mathbf{v} = \begin{matrix} 1 & \text{if } \mathbf{u} = \mathbf{v} \\ 0 & \text{if } \mathbf{u} \neq \mathbf{v} \end{matrix} \tag{4–71}$$

where the asterisk here denotes complex conjugation. Thus, the eigenvectors given are orthonormal.

In general, for a rotation in a rectilinear coordinate system, a real eigenvector corresponds to the direction of the rotation axis. If $\sin\theta = 0$, all the eigenvalues are real. If two eigenvalues are equal, any linear combination of the corresponding eigenvectors is an eigenvector. For each point symmetry operator there is a real eigenvalue whose eigenvector is in the direction of the symmetry axis; for a proper rotation axis this eigenvalue is $+1$, and for an improper rotation axis it is -1.

EXERCISE 4–46 Prove that if two eigenvalues of an operator **R** are equal, with eigenvectors **u** and **v**, the vector $a\mathbf{u} + b\mathbf{v}$ is also an eigenvector, where a and b are arbitrary coefficients.

EXERCISE 4–47 In a cartesian system, find the eigenvalues and eigenvectors of the operators

$$\begin{pmatrix} 1 & 0 & 0 \\ 0 & \bar{1} & 0 \\ 0 & 0 & 1 \end{pmatrix} \quad \text{and} \quad \begin{pmatrix} 1 & 0 & 0 \\ 0 & 1 & 0 \\ 0 & 0 & 1 \end{pmatrix}$$

EXERCISE 4–48 Find the eigenvalues and eigenvectors for

$$\begin{pmatrix} 0 & \bar{1} & 0 \\ 1 & 0 & 0 \\ 0 & 0 & 1 \end{pmatrix}$$ in a cartesian system. What is the angle θ?

EXERCISE 4–49 Find the eigenvalues and eigenvectors for

$$\begin{pmatrix} 0 & 1 & 0 \\ 0 & 0 & 1 \\ 1 & 0 & 0 \end{pmatrix}$$ in a cartesian system. What is the angle θ?

EXERCISE 4–50 Let $\mathbf{Q} = \mathbf{R}_x(90°)\mathbf{R}_y(120°)$. Find the eigenvalues of **Q** and the corresponding rotation axis and rotation angle. What does this result suggest about the possibility of having a fourfold axis perpendicular to a threefold axis in a finite group?

EXERCISE 4–51 Determine the rotation that is equivalent to the combination of two mutually perpendicular threefold rotation axes. Does such a combination of symmetry elements seem possible in a finite group?

EXERCISE 4-52 Find the real principal axis and the rotation angle of the operator

$$\begin{pmatrix} \bar{1} & 1 & 0 \\ \bar{1} & 0 & 0 \\ 0 & 0 & 1 \end{pmatrix}$$

What crystal system does this operator belong to?

EXERCISE 4-53 Find the principal axes and the rotation angle of the operator

$$\begin{pmatrix} 1 & \bar{1} & 0 \\ 0 & \bar{1} & 0 \\ 0 & 0 & \bar{1} \end{pmatrix}$$

EXERCISE 4-54 Find the symmetry operation and direction of the operator

$$\begin{pmatrix} 1 & 0 & 0 \\ 1 & \bar{1} & 0 \\ 0 & 0 & 1 \end{pmatrix}$$

EXERCISE 4-55 Find the principal axes and the rotation angle of

$$\begin{pmatrix} 1 & 0 & 0 \\ 1 & \bar{1} & 0 \\ 0 & 0 & \bar{1} \end{pmatrix}$$

EXERCISE 4-56 Find the principal axes and rotation angle of

$$\begin{pmatrix} 0 & 1 & 0 \\ 1 & 0 & 0 \\ 0 & 0 & 1 \end{pmatrix}$$

EXERCISE 4-57 Find the principal axes and rotation angle of

$$\begin{pmatrix} 0 & \bar{1} & 0 \\ 1 & 0 & 0 \\ 0 & 0 & \bar{1} \end{pmatrix}$$

EXERCISE 4-58 Find the principal axes and rotation angle of

$$\begin{pmatrix} 0 & 1 & 0 \\ 1 & 0 & 0 \\ 0 & 0 & \bar{1} \end{pmatrix}$$

CHAPTER 5

Tensor Properties of Crystals

5–1 Introduction

The state of a physical system may be described by specifying the values of a sufficient number of variables. Some of these variables, such as temperature, pressure, and volume, are conveniently expressed directly. In other cases, knowledge is more easily obtained of some externally imposed parameters, and the actual state variables are implied by mathematical relationships involving certain physical properties; an example here is an electric field that produces a state of polarization that is dependent upon the polarizability of the system.

If a single number suffices to denote the value of a variable at a point, the quantity is said to be a *scalar*; temperature is thus a scalar quantity. The temperature may change from point to point, in which case complete specification requires knowledge of how the temperature varies with direction, and the values of the temperature gradient along each of three independent directions constitute the components of a vector. Temperature gradient is thus an example of a vector property.

The next level of complexity introduces tensors of second or higher rank. We may consider, for example, a system subjected to a set of forces that are completely described by a second-rank stress tensor, in which each pair of indices indicates the direction of a force and the orientation of the surface upon which the force is acting. The stress tensor and the resulting deformation (represented by a strain tensor) will be examined in more detail in Sections 5–4, and 5–5.

Both the independent variables that may be controlled by the experimenter and the variables that describe the resulting state of the system may thus be regarded as tensor components (of zero rank for scalars, first rank for vectors, etc.). The mathematical relationship between the applied field tensor and the tensor expressing its effect is therefore a tensor equation. In many cases, such equations are linear in their tensors, at least as a first approximation. If a linear equation relates one tensor (say **A**) to another (say **B**), the coefficients in the equation are themselves tensors (say **C**) of appropriate rank. That is,

$$\mathbf{B} = \mathbf{CA} \tag{5–1}$$

Many physical laws are of this form. If, for example, **A** represents an electric field, **B** may represent current density, and **C** is then the electrical conductance tensor. Or if **A** is a stress tensor, **B** may be a strain tensor expressing the mechanical deformations of the system, and **C** is then the fourth-rank tensor of elastic compliances.

In general, if **A** represents an applied field and **B** expresses the consequences of the applied field in the system, **C** is a property of the system. Thus, a change in **A** leads to associated changes in **B**, but **C** depends only upon the system (and perhaps upon characteristics other than those represented by **A** or **B**). The physical property tensor **C** is sometimes called a *matter* tensor, while **A** and **B** are called *field* tensors. There is not necessarily a clearcut distinction as to which tensor, **A** or **B**, represents the cause and which represents the effect, and Eq. (5–1) may be solved for **A** to give

$$\mathbf{A} = \mathbf{C}^{-1}\mathbf{B}$$

where $\mathbf{C}^{-1}$ is a tensor that is at least symbolically the inverse of $\mathbf{C}$.

Dozens of books and thousands of articles have been published on the physical properties of crystals, and the brief treatment possible here can merely introduce the subject and use it as a source of further illustrations of the principles of tensor analysis. The following references are suggested to flesh out the skeletal development presented here. First, the book *Physical Properties of Crystals* by J. F. Nye (1957) is remarkably elegant, thorough, and lucid; the present work owes much to the illumination provided by Nye. The book *Tensor Properties of Materials* by A. R. Billings (1969) provides a good treatment of the elastic and transport properties of solids. An especially valuable reference on elasticity in generalized systems is *Tensor Analysis and Continuum Mechanics* by W. Flügge (1972). Concise and clear introductions may be found in the articles *Macroscopic Symmetry and Properties of Crystals*, by C. S. Smith (1958), and *Tensor Description of the Physical Properties of Crystals*, by T. K. Gaylord (1975).

Most problems in crystal physics are at least as tractable in cartesian coordinate systems as they are in crystal systems. Some equations are more compact or neater when referred to crystal coordinates, but the computational advantages are usually on the side of cartesian coordinates. Most treatments of crystal properties, Nye's book included, confine discussion to cartesian systems. Generalized treatments of such topics as stress, strain, and elasticity are, of course, essential in considerations of fluids or other materials where curvilinear geometry is induced by experimental constraints. We shall, in addition, find it helpful to generalize to at least rectilinear systems in order to illustrate the principles of tensor analysis developed in the preceding chapters.

The following list of tensor properties mentioned by Nye (1957) shows the range of topics susceptible to analysis by tensor methods. Gaylord (1975) gives an even more extensive list.

Diamagnetic susceptibility	Photoelasticity
Dielectric constants	Piezocaloric effect
Double refraction	Piezoelectricity
Elasticity	Pyroelectricity
Electrical conductivity	Pyromagnetism
Electric polarization	Strain

Electrocaloric effect	Stress
Electro-optical effects	Thermal conductivity
Ferroelectricity	Thermal expansion
Optical activity	Thermoelasticity
Paramagnetic susceptibility	Thermoelectricity

Some other references relevant to the content of this chapter are Barsch (1976) on x-ray determination of piezoelectric constants; Fumi and Ripamonti (1980) on tensor properties and rotational symmetry of crystals; Haussühl (1977) on elastic, thermoelastic, and piezoelectric properties of potassium bromate; the compilation of Huntington (1958) of elastic constants of crystals; and the handbook of elastic constants by Simmons and Wang (1971).

5-2 Second-Rank Tensor Properties

As a first example of a second-rank tensor property, consider electrical conductivity in a crystal. Suppose that an electric field **E** is applied to a crystal. Then

$$\mathbf{E} = E^i \mathbf{a}_i \tag{5-2}$$

where the E^i are the components of **E** along the crystal axes. The field may produce a current density **j**, with components j^i:

$$\mathbf{j} = j^i \mathbf{a}_i \tag{5-3}$$

The components of **j** and **E** are related by the generalization of Ohm's law,

$$j^i = \sigma^i{}_k E^k \tag{5-4}$$

where the $\sigma^i{}_k$ are the components of the electrical conductivity tensor. It is not imperative that we use contravariant components for both **j** and **E**; it is, in fact, often desirable to define **E** as the negative of the gradient of the potential (see, for example, McConnell, 1957)

$$E_i = -\frac{\partial \Phi}{\partial x^i} \tag{5-5}$$

and in terms of these covariant components Eq. (5–4) becomes

$$j^i = \sigma^{ik} E_k \tag{5-6}$$

In general, **j** and **E** are not parallel with each other, and the angle ψ between them is given by

$$\cos \psi = \frac{\sigma^{ik} E_i E_k}{jE} \tag{5-7}$$

If, however, **E** is applied in the direction of one of the principal axes of the tensor σ, so that

$$\mathbf{E} = K\mathbf{v} \tag{5-8}$$

where **v** is an eigenvector of σ and K is a constant, Eq. (5–6) may be written

$$j^i = \sigma^{ik} K v_k = \lambda K v^i = \lambda E^i \tag{5-9}$$

where λ is the eigenvalue belonging to **v**. Thus, **j** is parallel to **E** when **E** lies along a principal axis of σ, and the magnitude of **j** is just λ times the magnitude of **E**.

The electrical conductivity tensor is symmetric; that is,

$$\sigma^{ij} = \sigma^{ji} \tag{5-10}$$

It is not obvious that this should be so, and indeed all second-rank tensors are not symmetric. Proofs of the symmetry of second-rank tensors require thermodynamic arguments. When the tensor represents an equilibrium property (such as magnetic susceptibility or dielectric constants), the proof uses the laws of equilibrium thermodynamics and is relatively straightforward. Nonequilibrium properties (such as electrical or thermal conductance) require Onsager's theory of irreversible thermodynamics. If the conductance tensors were not symmetric, conduction in certain low-symmetry crystals would follow spiral paths. Long before the development of Onsager's theory, careful experiments were carried out to try to detect spiral heat flow; these experiments of course were in vain (see Miller, 1960). A characteristically lucid discussion of the symmetry properties of second-rank tensors is given by Nye; of all the second-rank tensors described by Nye, only the thermoelectric tensor is not symmetric.

According to Eq. (3–136), the magnitude of σ in the direction of a unit vector **u** is given by

$$\sigma(\mathbf{u}) = \tilde{\mathbf{u}}\sigma\mathbf{u} = u_i u_j \sigma^{ij} = u^i u^j \sigma_{ij}$$

$$= g_{ik} g_{jl} u^k u^l \sigma^{ij} \tag{5–11}$$

The magnitude of σ in the direction of an electric field **E** is equal to the length of the component of the current density **j** in the direction of **E** divided by the length of **E**.

The actual conductance $\varkappa$ of a sample in direction **u** is given by

$$\varkappa = \sigma(\mathbf{u})A/l \tag{5–12}$$

where A is the cross-sectional area and l is the length of the sample. The resistance R is

$$R = 1/\varkappa \tag{5–13}$$

It is useful also to define a resistance tensor ϱ, which is the reciprocal of σ:

$$\varrho = \sigma^{-1} \tag{5–14}$$

Note, however, that $\varrho(\mathbf{u})$ is not generally the reciprocal of $\sigma(\mathbf{u})$. In the experimental determination of $\sigma(\mathbf{u})$, **E** is in the direction of **u**, but **j** is not (unless **u** is an eigenvector), although only the component of **j** along **u** is important to the measurement. A corresponding experiment to measure $\varrho(\mathbf{u})$ would require **j** along **u** while permitting **E** to have other components.

Any physical property of a crystal must be invariant under the symmetry operations of the point group of the crystal. Otherwise, observation of the property could distinguish among what are supposed to be indistinguishable directions. This is known as Neumann's principle. By Eqs. (4–55) and (4–56)

$$\sigma = \mathbf{R}\sigma\overline{\mathbf{R}} \tag{5-15}$$

$$\sigma^{ij} = R^i{}_k \, \sigma^{kl} R^j{}_l \tag{5-16}$$

where $\mathbf{R}$ is a symmetry operator. Suppose, for example, that a crystal belongs to point group $C_3 - 3$. The symmetry operators referred to hexagonal axes are represented by the matrices $\mathbf{C}_3$ and $(\mathbf{C}_3)^2$ of Table 4-1 [see also Eq. (4-36)]. By Eq. (5-15)

$$\sigma = \mathbf{C}_3\sigma(\mathbf{C}_3)^2 = (\mathbf{C}_3)^2\sigma\mathbf{C}_3 \tag{5-17}$$

which leads to the relationships

$$\begin{aligned}
\sigma^{11} &= \sigma^{22} = \sigma^{11} - 2\sigma^{12} + \sigma^{22} \\
\sigma^{12} &= -\sigma^{12} + \sigma^{22} = \sigma^{11} - \sigma^{12} \\
\sigma^{13} &= -\sigma^{23} = -\sigma^{13} + \sigma^{23} \\
\sigma^{22} &= \sigma^{11} - 2\sigma^{12} + \sigma^{22} = \sigma^{11} \\
\sigma^{23} &= \sigma^{13} - \sigma^{23} = -\sigma^{13}
\end{aligned}$$

Therefore, σ has the form

$$\sigma = \begin{pmatrix} \sigma^{11} & \tfrac{1}{2}\sigma^{11} & 0 \\ \tfrac{1}{2}\sigma^{11} & \sigma^{11} & 0 \\ 0 & 0 & \sigma^{33} \end{pmatrix} \tag{5-18}$$

Transformation to a cartesian coordinate system by

$$\sigma' = \mathbf{F}\sigma\overline{\mathbf{F}} \tag{5-19}$$

with $\overline{\mathbf{F}}^{-1}$ given by Eq. (4-35) produces

$$\sigma' = \begin{pmatrix} \sigma'^{11} & 0 & 0 \\ 0 & \sigma'^{11} & 0 \\ 0 & 0 & \sigma'^{33} \end{pmatrix} \tag{5-20}$$

Table 5-1 shows the conditions imposed by symmetry upon the components of a symmetric second-rank tensor referred to crystal axes; Table 5-2 shows these conditions when the tensors are referred to cartesian axes.

TABLE 5-1. CONDITIONS IMPOSED BY SYMMETRY UPON THE COMPONENTS OF A SYMMETRIC SECOND-RANK TENSOR REFERRED TO CRYSTAL AXES

Crystal system	σ^{11}	σ^{12}	σ^{13}	σ^{22}	σ^{23}	σ^{33}
Triclinic	σ^{11}	σ^{12}	σ^{13}	σ^{22}	σ^{23}	σ^{33}
Monoclinic	σ^{11}	0	σ^{13}	σ^{22}	0	σ^{33}
Orthorhombic	σ^{11}	0	0	σ^{22}	0	σ^{33}
Tetragonal	σ^{11}	0	0	σ^{11}	0	σ^{33}
Hexagonal, trigonal	σ^{11}	$\frac{1}{2}\sigma^{11}$	0	σ^{11}	0	σ^{33}
Cubic	σ^{11}	0	0	σ^{11}	0	σ^{11}

TABLE 5-2. CONDITIONS IMPOSED BY SYMMETRY UPON THE COMPONENTS OF A SYMMETRIC SECOND-RANK TENSOR REFERRED TO CARTESIAN AXES

Crystal system	σ^{11}	σ^{12}	σ^{13}	σ^{22}	σ^{23}	σ^{33}
Triclinic	σ^{11}	σ^{12}	σ^{13}	σ^{22}	σ^{23}	σ^{33}
Monoclinic	σ^{11}	0	σ^{13}	σ^{22}	0	σ^{33}
Orthorhombic	σ^{11}	0	0	σ^{22}	0	σ^{33}
Tetragonal	σ^{11}	0	0	σ^{11}	0	σ^{33}
Hexagonal, trigonal	σ^{11}	0	0	σ^{11}	0	σ^{33}
Cubic	σ^{11}	0	0	σ^{11}	0	σ^{11}

EXERCISE 5-1 The specific resistance of bismuth metal at 20°C is 109×10^{-6} ohm-cm perpendicular to $\mathbf{a}_3$ and 138×10^{-6} ohm-cm parallel to $\mathbf{a}_3$. Bismuth is rhombohedral, space group $R\bar{3}m$, with $a_1 = 4.537$, $a_3 = 11.838$ Å for the hexagonal cell.

(a) Calculate the components of the conductance tensor referred to cartesian coordinates with $\mathbf{e}_3$ along $\mathbf{a}_3$.

(b) Calculate the components of the conductance tensor referred to the hexagonal axes.

(c) Calculate the resistance of a single crystal bismuth rod 3.00 cm long and 1.00 cm in diameter cut with the cylinder axis along [001].

(d) Suppose that a single crystal of bismuth is cut into a cylinder 3.00 cm long and 1.00 cm in diameter with the cylinder axis along the [211] direction (a rhombohedral axis). Calculate the resistance to the flow of current along [211] when an electric field is applied to the ends of the rod.

EXERCISE 5-2 Calculate the specific conductance of bismuth (see Exercise 5-1) along the hexagonal [211], $[\bar{1}11]$, and $[\bar{1}\bar{2}1]$ directions.

EXERCISE 5-3 Evaluate the average value of a second-rank cartesian tensor by evaluating $\int \bar{\mathbf{u}}\chi\mathbf{u}\, d\tau / \int d\tau$. Hint: Take $\bar{\mathbf{u}} = (\sin\theta \cos\phi, \sin\theta \sin\phi, \cos\theta)$, $d\tau = \sin\theta\, d\theta\, d\phi$, θ varies from 0 to π, ϕ varies from 0 to 2π.

EXERCISE 5-4 Magnetic susceptibility is another second-rank tensor property. Measurements on a zircon crystal gave the following values of the magnetic susceptibility. The values are referred to a set of arbitrarily chosen cartesian axes.

Direction	χ (rationalized mks units)
[100]	0.155
[010]	−0.0776
[001]	0.315
$[1/\sqrt{2}, 1/\sqrt{2}, 0]$	0.212
$[1/\sqrt{2}, 0, 1/\sqrt{2}]$	0.632
[111]	0.652

(a) Calculate all six components of the symmetric cartesian magnetic susceptibility tensor.

(b) Find the principal axes and principal values (eigenvalues) of this tensor.

(c) Zircon is tetragonal. How must the principal axes found in (b) be related to the tetragonal axes?

(d) In what direction in a zircon crystal will the magnetic susceptibility be zero?

(e) Positive values of the magnetic susceptibility correspond to paramagnetism, negative values to diamagnetism. Would polycrystalline zircon be paramagnetic or diamagnetic? What would be its susceptibility?

EXERCISE 5-5 The principal values A'^{11} and A'^{22} of a symmetric second-rank cartesian tensor of the form

$$\begin{pmatrix} A^{11} & A^{12} & 0 \\ A^{12} & A^{22} & 0 \\ 0 & 0 & A^{33} \end{pmatrix}$$

may be determined by the Mohr circle construction illustrated in Fig. 5-1, in which points are placed along the abscissa at distances from 0 equal to A^{11} and A^{22}, point C is midway between A^{11} and A^{22}, point R is at distance A^{12} from the larger of A^{11} and A^{22} (A^{22} in the figure) measured in a direction perpendicular to the A^{11}–A^{22} line. The circle centered at C and passing through R intersects the abscissa at A'^{11} and A'^{22}. Derive equations for A^{11}, A^{22}, and A^{12} in terms of the principal values A'^{11} and A'^{22} and the angle θ in the figure. Show that the result corresponds to rotation of the cartesian axes through θ about $\mathbf{e}_3$.

EXERCISE 5-6 The thermal conductivity of single-crystal bismuth at 25°C is 0.0530 watt/deg cm parallel to the hexagonal $\mathbf{a}_3$ axis and 0.0919 watt/deg cm normal to $\mathbf{a}_3$ (Ho, Powell, and Liley, 1972). See Exercise 5-1 for the unit cell dimensions.

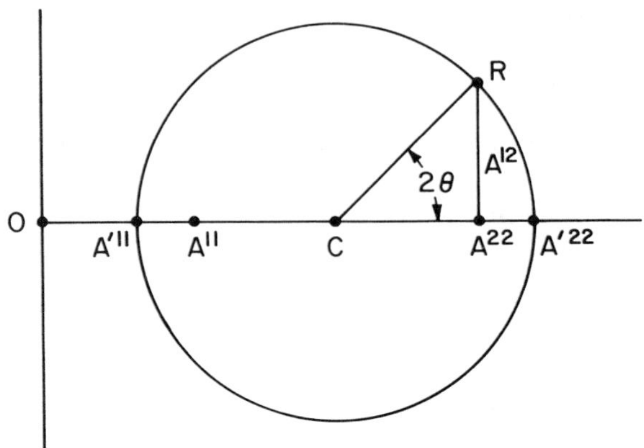

FIGURE 5-1. Mohr circle construction (see Exercise 5-5).

(a) Calculate the thermal conductivity of a single crystal of bismuth along the hexagonal [211] and [101] directions.

(b) Suppose that a temperature difference of 25°C is maintained across the ends of the crystal of Exercise 5-1(c). Calculate the rate at which heat is transferred through the crystal.

EXERCISE 5-7 Carslaw and Jaeger (1947) give the thermal conductivity of spruce wood as 5.5×10^{-4} cal/cm-deg-sec with the grain and 3.0×10^{-4} cal/cm-deg-sec across the grain.

(a) Calculate the thermal conductivity of spruce in a direction 30° to the grain.

(b) Calculate the average thermal conductivity of spruce.

5-3 Representation Quadrics

Associated with a second-rank tensor, such as σ, is a surface, known as the representation quadric, that is generated by a vector **h** that satisfies the equation

$$\mathbf{h}\sigma\mathbf{h} = 1 \qquad (5\text{-}21)$$

or

$$h_i h_j \sigma^{ij} = 1 \qquad (5\text{-}22)$$

When referred to the principal axes $\mathbf{h}'$ of σ, the equation of this surface may be written

$$(h_1')^2\sigma'^{11} + (h_2')^2\sigma'^{22} + (h_3')^2\sigma'^{33} = 1 \qquad (5\text{-}23)$$

where the σ'^{II} are the principal values of σ. If the σ'^{II} are all positive, the surface is an ellipsoid. If one of the values of σ'^{II} is negative, the surface is a hyperboloid of one sheet. If two of the σ'^{II} values are negative, the surface is a hyperboloid of two sheets. If all three principal values are negative, the surface may be regarded as an imaginary ellipsoid, or it may be more convenient to replace 1 by -1 in Eq. (5–21) and represent the property by a real ellipsoid.

Many of the second-rank tensors of physical interest are positive-definite (all eigenvalues positive), so that their representation quadrics are ellipsoids with semi-axes of lengths $1/\sqrt{\sigma'^{II}}$. This reciprocal relationship should be kept in mind; a short ellipsoid axis corresponds to a large value of the physical property, and vice versa. Suppose that $\mathbf{v}$ is a vector drawn from the center of the ellipsoid to a point on the surface. Writing

$$\mathbf{v} = v\mathbf{u} \qquad (5\text{-}24)$$

where $\mathbf{u}$ is a unit vector in the direction of $\mathbf{v}$, $\mathbf{v}$ must satisfy

$$\bar{\mathbf{v}}\sigma\mathbf{v} = 1 \qquad (5\text{-}25)$$

Hence,

$$(v)^2\bar{\mathbf{u}}\sigma\mathbf{u} = 1 \qquad (5\text{-}26)$$

But $\bar{\mathbf{u}}\sigma\mathbf{u}$ is the magnitude of σ in the direction of $\mathbf{u}$. Thus, $(v)^2$ is equal to the reciprocal of the magnitude of the property in the direction of $\mathbf{v}$.

Another surface of interest is the ellipsoid generated by the vector

$$\mathbf{t} = \sigma\mathbf{u} \qquad (5\text{-}27)$$

where $\mathbf{u}$ is a unit vector. Solving Eq. (5–27) for $\mathbf{u}$ and setting $\bar{\mathbf{u}}\mathbf{u}$ equal to unity gives

$$\bar{\mathbf{t}}\sigma^{-1}\sigma^{-1}\mathbf{t} = 1 \tag{5-28}$$

If referred to a cartesian system aligned along the principal axes, Eq. (5-28) becomes (with primes denoting principal values)

$$\frac{(t^1)^2}{(\sigma'^{11})^2} + \frac{(t^2)^2}{(\sigma'^{22})^2} + \frac{(t^3)^2}{(\sigma'^{33})^2} = 1 \tag{5-29}$$

which is the equation of an ellipsoid whose semi-axes have lengths σ'^{II}. The surface, of course, remains an ellipsoid upon transformation of axes, so Eq. (5-28) is the equation of an ellipsoid. By arguments similar to those applied to Eq. (5-26), it follows that $(t)^2$ equals the reciprocal of the magnitude of σ^{-2} in the direction of $\mathbf{t}$. The ellipsoid generated by $\mathbf{t}$ is called the *magnitude ellipsoid.*

EXERCISE 5-8 Describe the surface generated by Eq. (5-23) for the magnetic susceptibility of zircon described in Exercise 5-4.

EXERCISE 5-9 Describe both the representation surface [given by Eq. (5-23)] and the magnitude ellipsoid of the thermal conductivity of spruce wood (Exercise 5-7).

5-4 Stress

Forces acting upon a physical system may be categorized as *body forces*, such as gravitation, which act upon each particle of the system, or as *contact forces*, which act upon the surface of each volume element. The contact forces acting upon a volume element may be described in terms of a second-rank tensor, in which each pair of indices denotes the direction of a force and the orientation of the surface upon which the force acts. If the components are chosen to be forces per unit area, the tensor is called a *stress tensor*. We shall denote the stress tensor by the symbol σ (not to be confused with electrical conductance, for which the same symbol was used). A proof that σ is a tensor may be found in Nye (1957).

The stress tensor represents the set of forces that are imposed upon a system. It is an example of a *field tensor*, which was mentioned in Section 5-1. Field tensors do not represent properties of the system, and they do not have to conform to the symmetry of the system. A compressive stress might, for instance, be applied along $\mathbf{a}_1$ of a cubic crystal, with no stress whatsoever along the crystallographically equivalent $\mathbf{a}_2$ and $\mathbf{a}_3$.

It will be adequate for our purposes to refer stress tensors to cartesian coordinate systems. Considering a rectangular volume element aligned with the cartesian axes, component σ^{ij} is defined as the force per unit area acting in the $\mathbf{e}_i$ direction upon the surface normal to $\mathbf{e}_j$. Figure 5-2 depicts the stress components acting upon the surface normal to $\mathbf{e}_3$.

The quantities σ^{11}, σ^{22}, and σ^{33} are the *normal* components of stress; Nye follows the convention of assigning positive values to *tensile* (or stretching) stresses, and negative values to *compressive* stresses. The components σ^{ij} with $i \neq j$ are the *shear components.* The nature of shear component σ^{23} in Fig. 5-2 may be understood by imagining that the cube is held between parallel plates applied to the surfaces normal to $\mathbf{e}_3$; if A is the area of each of these surfaces, a force $\sigma^{23}A$ pulls the upper face to the right, and an equal and opposite force pulls the lower face to the left.

A *homogeneous* stress consists of forces that are independent of the location of the volume element in the system. For a homogeneous stress, the moment tending to rotate the volume element of Fig. 5-2 clockwise about $\mathbf{e}_1$ is $\sigma^{23}abc/2$, where ab is the area of the face nor-

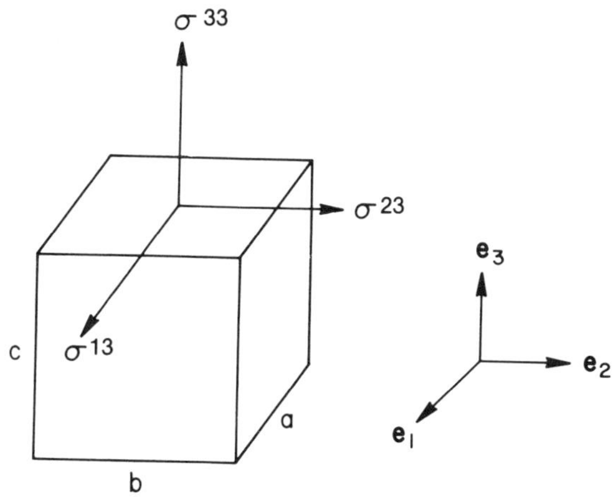

FIGURE 5-2. Stress components σ^{i3}.

mal to $\mathbf{e}_3$, and $c/2$ is the moment arm to the center of the volume element. The moment tending to produce counterclockwise rotation about $\mathbf{e}_1$ is $\sigma^{32}acb/2$, where ac is the area of the face upon which σ^{32} acts and $b/2$ is the corresponding moment arm. The condition for no net rotation is thus that σ^{23} and σ^{32} are equal. This result may be generalized to

$$\sigma^{ij} = \sigma^{ji} \tag{5-30}$$

Nye gives a proof that Eq. (5-30) is valid also when the stress is inhomogeneous, provided that there are no body torques. Thus, σ is a symmetric second-rank tensor.

EXERCISE 5-10 A cube 2.00 cm on an edge is placed in a press that exerts a force of 7.50 newtons in the $\mathbf{e}_3$ direction. The cube edges are parallel to the cartesian axes. Write all components of the stress tensor.

EXERCISE 5-11 A mass of 60.0 grams is hung from a weightless rod of cross section 5.00 cm². The acceleration due to gravity is 970 cm/sec². Write all components of the stress tensor.

EXERCISE 5-12 Suppose that the only nonzero elements of a stress tensor are $\sigma^{12} = \sigma^{21}$. Find the principal values of σ, thus showing that a shear can be resolved into a stretch and a compression.

EXERCISE 5-13 A hydrostatic stress in a cartesian system is defined by $\sigma^{ij} = -p\delta^{ij}$, where p is the hydrostatic pressure and δ^{ij} is the Kronecker delta. Calculate the tensor components of the hydrostatic stress in a generalized rectilinear coordinate system.

5-5 Strain

A crystal (or any other system) subjected to stress may deform. As an externally imposed stress need not possess the symmetry of the crystal, neither is the deformation constrained by the crystal symmetry. Thus, neither stress nor the resulting deformation is a crystal property. The deformation may, in fact, alter the symmetry of a crystal, and possibly even change the crystal system.

The treatment of deformation is facilitated by choosing basis vectors that connect particular particles or atoms; as the system is deformed, the atoms move and hence the basis vectors change relative to their original directions and lengths. If the system is a crystal, the natural basis vectors for the undeformed state are lattice vectors. The translational equivalence of lattice points may be destroyed by such deformations as bending and twisting; hence, even in the crystalline state it is advantageous to specify each basis vector as the vector connecting a particular pair of atoms.

A measure of the deformation is given by

$$\epsilon = \tfrac{1}{2}(\mathbf{g}' - \mathbf{g}) \tag{5-31}$$

where $\mathbf{g}$ and $\mathbf{g}'$ are the metric tensors before and after deformation, respectively, and ϵ is the *strain tensor* (not to be confused with the permutation tensor, for which the same symbol was used). In suffix notation

$$\epsilon_{ij} = \tfrac{1}{2}(g'_{ij} - g_{ij}) \tag{5-32}$$

It is apparent from the symmetry of the metric tensors that ϵ is symmetric. The cases we shall consider are those of *homogeneous strain*, in which ϵ is the same at every point of the system.

Some justification is needed of the blithe reference to ϵ as a tensor. After all, ϵ is defined as the difference of two metric tensors, which specify the metric properties of two coordinate systems, and there would seem to be some ambiguity about which set of basis vectors ϵ is referred to. We shall choose to refer ϵ to the original undeformed axes, and we may then think of the elements of $\mathbf{g}'$ as being the scalar products of a triplet of vectors defined in the undeformed system. Thus, $\mathbf{g}$, $\mathbf{g}'$, and hence ϵ are all doubly covariant tensors relative to the original axes.

The relationship between the original axes $\mathbf{a}_i$ in the undeformed system and the axes $\mathbf{a}'_i$ after deformation may be written

$$\mathbf{a}' = \mathbf{G}\mathbf{a} \tag{5-33}$$

or

$$\mathbf{a}'_i = G_i{}^j\,\mathbf{a}_j \tag{5-34}$$

It is important to observe the essential difference between the axial relationships indicated by Eqs. (5–33) and (5–34) and the usual transformations that were developed in Chapter 3. In the present case the atoms move relative to the original axes, whereas our previous transformations dealt with the selection of new axes when the atoms were fixed in space. The coordinates x'^i, referred to axes $\mathbf{a}_i'$, of an atom in the deformed crystal are the same numbers as the coordinates x^i referred to axes $\mathbf{a}_i$ of the atom in the undeformed crystal; this follows from the definition of the basis vectors in terms of specific atoms, so that an atom at say [100] in the undeformed $\mathbf{a}_i$ system will be at [100] in the deformed $\mathbf{a}_i'$ system. When the deformed coordinates x'^i are referred to the $\mathbf{a}_i$ axes they are given by $\mathbf{F}^{-1}\mathbf{x}'$, which of course is equal to $\overline{\mathbf{G}}\mathbf{x}'$. The coordinate shift is therefore

$$\mathbf{u} = \overline{\mathbf{G}}\mathbf{x}' - \mathbf{x} \tag{5–35}$$

or

$$u^i = G_j^{\ i} x'^j - x^i \tag{5–36}$$

Since x^i and x'^i are equal,

$$u^i = (G_j^{\ i} - \delta_j^{\ i})x^j \tag{5–37}$$

The covariant components of $\mathbf{u}$ are

$$u_i = g_{ik}(G_j^{\ k} - \delta_j^{\ k})x^j \tag{5–38}$$

We now define a matrix $\mathbf{U}$ with elements

$$U_{in} = \frac{\partial u_i}{\partial x^n} \tag{5–39}$$

In a rectilinear system g_{ik} and $G_j^{\ k}$ are constants, and

$$U_{in} = g_{ik}(G_j^{\ k} - \delta_j^{\ k})\delta_n^{\ j} \tag{5–40}$$

$$U_{in} = g_{ik}(G_n^{\ k} - \delta_n^{\ k}) \tag{5–41}$$

In matrix form

$$\mathbf{U} = \mathbf{g}(\overline{\mathbf{G}} - \mathbf{I}) \tag{5-42}$$

It is easily verified that

$$\mathbf{U} + \overline{\mathbf{U}} + \overline{\mathbf{U}}\mathbf{g}^*\mathbf{U} = \mathbf{G}\mathbf{g}\overline{\mathbf{G}} - \mathbf{g} \tag{5-43}$$

Since

$$\mathbf{G}\mathbf{g}\overline{\mathbf{G}} = \mathbf{g}' \tag{5-44}$$

strain in rectilinear coordinate systems may be defined by (cf. Flügge, 1972)

$$\epsilon = \tfrac{1}{2}(\mathbf{U} + \overline{\mathbf{U}} + \overline{\mathbf{U}}\mathbf{g}^*\mathbf{U}) \tag{5-45}$$

$$\epsilon_{ij} = \frac{1}{2}\left(\frac{\partial u_i}{\partial x^j} + \frac{\partial u_j}{\partial x^i} + \frac{\partial u_k}{\partial x^j}\frac{\partial u^k}{\partial x^i}\right) \tag{5-46}$$

In curvilinear systems g_{ik} and $G_j{}^k$ are not constants, and the derivatives of coordinates are not tensors. Generalization of Eqs. (5–45) and (5–46) to curvilinear coordinate systems is given by Flügge (1972).

If the strain is very small, the third term on the right of Eq. (5–45) [or (5–46)] is negligible, and the strain is simply

$$\epsilon = \tfrac{1}{2}(\mathbf{U} + \overline{\mathbf{U}}) \tag{5-47}$$

The antisymmetric tensor

$$\omega = \tfrac{1}{2}(\mathbf{U} - \overline{\mathbf{U}}) \tag{5-48}$$

expresses the rotation of the system. If a system is subjected to a stress, the infinitesimal displacements produced may be resolved into the symmetric strain tensor ϵ, which describes the deformations within the system, and the antisymmetric rotation tensor ω, which expresses any total rotation of the entire system.

EXERCISE 5–14 A system is deformed so that the cartesian displacements are $u^1 = x^2 \sin\alpha$, $u^2 = 0$, $u^3 = 0$.

(a) Calculate the strain components from Eq. (5–46).

(b) Determine **G** from Eq. (5–42), and then find **g**′ from Eq. (5–44), and finally obtain the strain components from Eq. (5–32)

EXERCISE 5–15 A monoclinic crystal has $a_1 = 6.00$, $a_2 = 8.00$, $a_3 = 7.50$Å, $\beta = 103.0°$. The crystal is deformed so that $\mathbf{a}_1' = \mathbf{a}_1 + \mathbf{a}_3 \sin\phi$, $\mathbf{a}_2' = \mathbf{a}_2$, $\mathbf{a}_3' = \mathbf{a}_1 \sin\phi + \mathbf{a}_3$.

(a) Calculate **g**′ and determine the stress tensor referred to monoclinic axes by means of Eq. (5–32).

(b) For small ϕ, $\sin\phi \cong \phi$, $\phi^2 \ll \phi$. Apply this approximation to the result of (a) to obtain ϵ for small ϕ.

(c) Calculate the components of **u** and the elements of the matrix **U** and, using the approximation of small ϕ, determine ϵ from Eq. (5–45).

(d) Transform ϵ to a cartesian coordinate system chosen by $\mathbf{e}_1 = \mathbf{a}_1/a_1$, $\mathbf{e}_2 = \mathbf{a}_2/a_2$, $\mathbf{e}_3 = -\mathbf{a}_1 \cos\beta/(a_1 \sin\beta) + \mathbf{a}_3/(a_3 \sin\beta)$.

(e) Find the principal values of the cartesian strain tensor of (d).

EXERCISE 5–16 For a pure shear about $\mathbf{e}_3$ in a cartesian coordinate system, the only nonvanishing components of ϵ are $\epsilon_{12} = \epsilon_{21}$. Find the principal values and principal axes of ϵ.

EXERCISE 5–17 (a) Show that the change in the unit cell volume of a crystal subjected to a small deformation is $\Delta V = V g^{ij}\epsilon_{ij}$.

(b) Consider a crystal with

$$\mathbf{g} = \begin{pmatrix} 88.000 & 24.000 & 0 \\ 24.000 & 65.000 & 0 \\ 0 & 0 & 100.00 \end{pmatrix}$$

$$\epsilon = \begin{pmatrix} 0.035 & -0.040 & 0 \\ -0.040 & 0.035 & 0 \\ 0 & 0 & 0 \end{pmatrix}$$

Calculate $\mathbf{g}^*$, $\mathbf{g}'$, V, and V' and confirm numerically that $V' - V$ is given by the result of (a).

(c) Show that in a cartesian coordinate system $\Delta V = \text{trace}(\epsilon)$.

5–6 Thermal Expansion

An important special case of strain is that induced by a temperature change. The thermal expansion tensor α is defined by

$$\alpha_{ij} = \frac{d\epsilon_{ij}}{dT} \tag{5–49}$$

where T is the temperature.

The thermal expansion tensor of a crystal represents a crystal property and must, therefore, have all the symmetry of the crystal.

In terms of the metric tensor [see Eq. 5–31)]

$$\alpha_{ij} = \frac{1}{2}\frac{dg_{ij}}{dT} \tag{5–50}$$

$$\alpha_{ij} = \frac{1}{2}\left(\mathbf{a}_i \cdot \frac{d\mathbf{a}_j}{dT} + \mathbf{a}_j \cdot \frac{d\mathbf{a}_i}{dT}\right) \tag{5–51}$$

Thermal expansion coefficients that have dimensions of T^{-1} may be created by raising an index of α

$$\alpha^i{}_j = g^{ik}\alpha_{kj} \tag{5–52}$$

Thermal expansion data are given usually in terms of cartesian coordinate systems where the $\alpha^i{}_j$ and α_{ij} are equal numerically, but for dimensional consistency such values should be interpreted as components of the mixed tensor.

The change of unit cell volume with temperature is given by

$$\frac{dV}{dT} = \frac{d}{dT}(\mathbf{a}_1 \cdot \mathbf{a}_2 \Lambda \mathbf{a}_3) \tag{5–53}$$

which may be converted to

$$\frac{dV}{dT} = V\frac{d\mathbf{a}_i}{dT} \cdot \mathbf{a}^i \tag{5–54}$$

$$\frac{dV}{dT} = V\frac{d\mathbf{a}_i}{dT} \cdot (g^{ij}\mathbf{a}_j) \tag{5–55}$$

$$\frac{dV}{dT} = \frac{1}{2}V g^{ij}\frac{dg_{ij}}{dT} \tag{5–56}$$

$$\frac{dV}{dT} = Vg^{ij}\alpha_{ij} \tag{5–57}$$

The coefficient of volume expansion is

$$\frac{1}{V}\frac{dV}{dT} = g^{ij}\alpha_{ij} \qquad (5\text{–}58)$$

In cartesian coordinates this is

$$\frac{1}{V}\frac{dV}{dT} = \alpha_{ii} \qquad (5\text{–}59)$$

Weigel *et al.* (1978) have defined an aspherism index A that expresses the deviation of the thermal expansion tensor from isotropy in terms of the invariants i_1 and i_2, which in cartesian coordinates are (see Exercise 3–18)

$$i_1 = \text{trace}(\alpha) = \alpha_{ii} \qquad (5\text{–}60)$$

$$i_2 = |\alpha| \, \text{trace}(\alpha^{-1}) \qquad (5\text{–}61)$$

If α is referred to its principal cartesian axes,

$$i_2 = \alpha_{11}\alpha_{22} + \alpha_{22}\alpha_{33} + \alpha_{33}\alpha_{11} \qquad (5\text{–}62)$$

We now define

$$x_1 = \alpha_{11}/\alpha_{ii}, \qquad x_2 = \alpha_{22}/\alpha_{ii}, \qquad x_3 = \alpha_{33}/\alpha_{ii} \qquad (5\text{–}63)$$

so that

$$x_1 + x_2 + x_3 = 1 \qquad (5\text{–}64)$$

$$i_2 = (i_1)^2 (x_1x_2 + x_2x_3 + x_3x_1) \qquad (5\text{–}65)$$

In Fig. 5–3 a point P is located at distance x_i from the side opposite vertex i of an equilateral triangle of unit altitude. A property of equilateral triangles is that the sum of the distances x_i is equal to the altitude. The aspherism index A is the distance from P to the center C of the triangle. By elementary geometry

$$A = PC = \frac{2}{3}[3(x_1)^2 + 3(x_2)^2 - 3x_1 - 3x_2 + 3x_1x_2 + 1]^{1/2} \qquad (5\text{–}66)$$

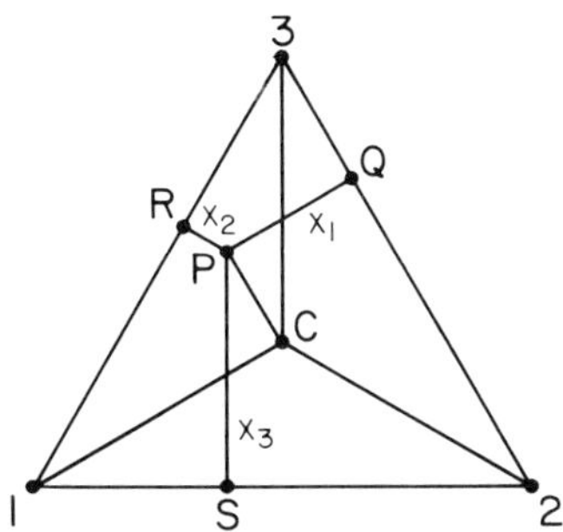

FIGURE 5-3 Asphericism index $A = PC$.

Therefore,

$$A = \frac{2}{3}[1 - 3i_2/(i_1)^2]^{1/2} \tag{5-67}$$

In a cubic crystal, $\alpha_{11} = \alpha_{22} = \alpha_{33}$, $i_1 = 3\alpha_{11}$, $i_2 = 3(\alpha_{11})^2$, and $A = 0$, and the thermal expansion is isotropic.

EXERCISE 5-18 Derive Eq. (5-54) from Eq. (5-53) and Eq. (5-56) from Eq. (5-54).

EXERCISE 5-19 The unit cell dimensions of $\alpha-NH_4HgCl_3$ (J. Sadanandan and S. V. V. Suryanarayana, 1979) are given by

$$a_1 = 4.1920 + 1.8413 \times 10^{-4}T$$
$$a_3 = 7.9319 + 1.6588 \times 10^{-4}T$$

where T is the absolute temperature, and the lengths are in angstroms. The crystal system is tetragonal.

(a) Calculate the components α_{ij} of the thermal expansion tensor in the tetragonal system. (Assume that the directions of the vectors do not change.)

(b) Transform α to a cartesian coordinate system aligned along the principal axes.

(c) Calculate the thermal expansion along the cartesian [110], [001], and [111] directions at 25°C.

EXERCISE 5-20 Derive Eq. (5-66) and Eq. 5-67).

EXERCISE 5-21 Show that for a tetragonal or hexagonal crystal (for which $\alpha_{11} = \alpha_{22}$),

$$A = \frac{2(\alpha_{11} - \alpha_{33})}{3(2\alpha_{11} + \alpha_{33})}$$

EXERCISE 5-22 According to T. Beguemsi, P. Garnier, and D. Weigel (1978), the principal values referred to cartesian axes of the thermal expansion tensor of orthorhombic V_2O_5 are $\alpha_{11} = -2.3 \times 10^{-6}\ K^{-1}$, $\alpha_{22} = 52.7 \times 10^{-6}\ K^{-1}$, and $\alpha_{33} = -2.2 \times 10^{-6}\ K^{-1}$ at 300°K. The corresponding unit cell dimensions are $a_1 = 11.51$, $a_2 = 4.37$, $a_3 = 3.565$ Å.

(a) Calculate i_1, i_2, and the aspherism index.

(b) How would you plot the negative values on the triangular diagram of Fig. 5-3?

(c) Calculate the coefficient of volume expansion.

5-7 Elasticity

When a stress, represented by tensor σ, is applied to a crystal, a deformation, represented by tensor ϵ, is produced. For small deformations, strain is related linearly to stress by Hooke's law, which in generalized form is

$$\epsilon = \mathbf{s}\sigma \tag{5-68}$$

where **s** is the elastic compliance tensor. In suffix form

$$\epsilon_{ij} = s_{ijkl}\sigma^{kl} \tag{5-69}$$

Thus, **s** is a fourth-rank covariant tensor, which represents a property of the system. The requirement that both σ and ϵ be symmetric makes impossible separate determination of the components s_{ijkl}, s_{jikl}, s_{ijlk}, and s_{jilk}, so no experimental information will be sacrificed by equating these components:

$$s_{ijkl} = s_{jikl} = s_{ijlk} = s_{jilk} \tag{5-70}$$

Also, $s_{ijkl} = s_{klij}$, so the number of independent components of **s** that a crystal can have is reduced from $(3)^4 = 81$ to 21.

The inverse of Eq. (5-68) is

$$\sigma = \mathbf{c}\epsilon \tag{5-71}$$

$$\sigma^{ij} = c^{ijkl}\epsilon_{kl} \tag{5-72}$$

where **c** is the fourth-rank contravariant elastic stiffness tensor. Note that the symbols **s** and **c** for the compliance and stiffness tensors are just the opposite of their initials. The stiffness tensor components follow the symmetry relationships

$$c^{ijkl} = c^{jikl} = c^{ijlk} = c^{jilk} = c^{klij}, \text{ etc.} \tag{5–73}$$

so there are at most 21 independent components of **c.**

By the rules developed in Chapter 3, the transformation properties of these fourth-rank tensors are

$$c'^{\,ijkl} = F^i{}_m\, F^j{}_n\, F^k{}_o\, F^l{}_p\, c^{mnop} \tag{5–74}$$

$$s'_{\,ijkl} = G_i{}^m\, G_j{}^n\, G_k{}^o\, G_l{}^p\, s_{mnop} \tag{5–75}$$

The invariance of these under the operations **R** of the point group is expressed by

$$c^{ijkl} = R^i{}_m\, R^j{}_n\, R^k{}_o\, R^l{}_p\, c^{mnop} \tag{5–76}$$

$$s_{ijkl} = R^*{}_i{}^m\, R^*{}_j{}^n\, R^*{}_k{}^o\, R^*{}_l{}^p\, s_{mnop} \tag{5–77}$$

where **R*** is the symmetry operator referred to the reciprocal axes.

Transformation of fourth-rank tensors tends to be cumbersome, and it is nearly always sufficient (and preferable) to use cartesian axes in the treatment of elasticity in crystals. The general theory referred to curvilinear axes may, however, be advantageous in some applications, such as those involving fluid flow (cf. Aris, 1962).

Another labor-saving device commonly used in treating elasticity is to introduce a two-index notation that condenses pairs of indices as follows

$$11 \rightarrow 1,\ 22 \rightarrow 2,\ 33 \rightarrow 3,\ 23 \rightarrow 4,\ 13 \rightarrow 5,\ 12 \rightarrow 6 \tag{5–78}$$

Instead of writing c^{ijkl} we thus write c^{mn}, where m is the single index corresponding to the pair ij and n is the single index corresponding to the pair kl. Thus, for example,

$$c^{1122} = c^{12}, \qquad c^{1233} = c^{63} = c^{36}, \qquad c^{2323} = c^{44} \tag{5–79}$$

The reduced-index analogue of Eq. (5–72) is

$$\sigma^i = c^{ij}\epsilon_j \tag{5–80}$$

where i and j both run from 1 to 6. The components σ^i are related directly to the components σ^{mn}

$$\begin{aligned} \sigma^{11} &= \sigma^1, & \sigma^{22} &= \sigma^2, & \sigma^{33} &= \sigma^3, \\ \sigma^{12} &= \sigma^6, & \sigma^{23} &= \sigma^4, & \sigma^{13} &= \sigma^5 \end{aligned} \tag{5-81}$$

If the correspondences of $\mathbf{c}$ and σ are defined in this way, some factors of ½ are needed in the correspondences of ϵ [otherwise, Eq. (5-80) will not be equivalent to Eq. (5-72)]; we thus write

$$\epsilon_{ij} = A\epsilon_m \tag{5-82}$$

where

$$A = 1 \qquad \text{if } m = 1, 2, \text{ or } 3$$

$$A = \tfrac{1}{2} \qquad \text{if } m = 4, 5, \text{ or } 6$$

In order to be able to use the reduced-index analogue of Eq. (5-69)

$$\epsilon_i = s_{ij}\sigma^j \tag{5-83}$$

it is necessary to introduce factors of ½ or ¼ in the correspondences of $\mathbf{s}$. We write

$$s_{ijkl} = Bs_{mn} \tag{5-84}$$

where

$B = 1$ if m and n are both 1, 2, or 3

$B = \tfrac{1}{2}$ if one of m or n is 4, 5, or 6 and the other is 1, 2, or 3

$B = \tfrac{1}{4}$ if both m and n are 4, 5, or 6

Thus, for example,

$$\epsilon_{11} = \epsilon_1, \qquad \epsilon_{23} = \tfrac{1}{2}\epsilon_4 \tag{5-85}$$

$$s_{1122} = s_{12}, \qquad s_{1233} = \tfrac{1}{2}s_{63} = \tfrac{1}{2}s_{36}, \qquad s_{2323} = \tfrac{1}{4}s_{44} \tag{5-86}$$

Equations (5–80) and (5–83) are now matrix equations where the indices run from 1 to 6; ϵ and σ are 6×1 matrices, and **c** and **s** are 6×6 matrices, which are related by

$$\mathbf{s} = \mathbf{c}^{-1}, \qquad \mathbf{c} = \mathbf{s}^{-1} \tag{5–87}$$

The relationships imposed by symmetry on cartesian stiffness and compliance tensors are summarized in Table 5–3.

EXERCISE 5–23 The compliances and stiffnesses of an elastically isotropic material obey all the symmetry conditions of the cubic crystal system plus invariance under rotation by 45° about each cartesian axis. Show that a 45° rotation about $\mathbf{e}_3$ leads to the relationship $c^{66} = \frac{1}{2}(c^{11} - c^{12})$.

EXERCISE 5–24 Show that c^{44} for an isotropic material is invariant under any rotation about $\mathbf{e}_3$. This result may be generalized to any component of **c** or **s** about any axis.

EXERCISE 5–25 Write out all nine terms of Eq. (5–69) for the particular case of ϵ_{12}. Replace ϵ_{12} by $\frac{1}{2}\epsilon_6$ and each s_{ijkl} by the appropriate s_{mn} from Eq. (5–86), and confirm the validity for this case of Eq. (5–83).

TABLE 5–3 NONZERO COMPONENTS OF s_{ij} AND c^{ij} FOR CARTESIAN AXES (The factor m is equal to 2 for s_{ij}, ½ for c^{ij}; the factor n is equal to ½ for s_{ij}, 1 for c^{ij}.)

Symmetry	Nonzero components and relationships
$1, \bar{1}$	No restrictions
$2, m, 2/m$	11, 12, 13, 15, 22, 23, 25, 33, 35, 44, 46, 55, 66
$222, mm2, mmm$	11, 12, 13, 22, 23, 33, 44, 55, 66
$4, \bar{4}, 4/m$	11 = 22, 12, 13 = 23, 16 = −26, 33, 44 = 55, 66
$4mm, \bar{4}2m, 422, 4/mmm$	11 = 22, 12, 13 = 23, 33, 44 = 55, 66
$3, \bar{3}$	11 = 22, 12, 13 = 23, 14 = −24 = $n \times$ 56, 15 = − 25 = $-n \times$46, 33, 44 = 55, 66 = $m \times (11-12)$
$32, 3m, \bar{3}m$	11 = 22, 12, 13 = 23, 14 = −24 = $n \times$56, 33, 44 = 55, 66 = $m \times (11-12)$
$6, \bar{6}, 6/m, 622, 6mm, \bar{6}m2, 6/mmm$	11 = 22, 12, 13 = 23, 33, 44 = 55, 66 = $m \times (11-12)$
$23, \bar{4}3m, m3, 432, m3m$	11 = 22 = 33, 12 = 13 = 23, 44 = 55 = 66
Isotropic	11 = 22 = 33, 12 = 13 = 23, 44 = 55 = 66 = $m \times (11-12)$

EXERCISE 5–26 Find the conditions imposed by symmetry upon the components c^{ijkl} referred to hexagonal axes for a trigonal crystal with symmetry $C_3 - \mathbf{3}$. If a computer is not available to assist with the determination of all the relationships, the principles may be illustrated sufficiently by writing out the relationships for symmetry operator **3** for c^{11}, for c^{43}, for c^{35}, and for c^{13}.

EXERCISE 5–27 Repeat Exercise 5–26 in a cartesian coordinate system.

EXERCISE 5–28 Transform the tensor components c^{ijkl} of a trigonal crystal (see Exercise 5–26) to a cartesian coordinate system and show the relationships between the trigonal c^{ij} and the cartesian c'^{ij}.

EXERCISE 5–29 (a) Show that $s_{ijkl}c^{klmn} = \delta_i^m \delta_j^n$ and $c^{klmn} s_{mnop} = \delta^k_o \delta^l_p$

(b) Write out all six terms of $s_{4q}c^{q3}$, replace each s_{pq} and c^{rs} by its four-index equivalent, and confirm that the equations derived in (a) agree with the 6×6 matrix equation $\mathbf{cs} = \mathbf{I}$.

EXERCISE 5–30 The acoustic gyrotropic tensor is a fifth-rank tensor (D. L. Portigal and E. Burstein, 1968; K. Kumaraswamy and N. Krishnamurthy, 1980).

(a) What is the total number of components possible for a fifth-rank tensor?

(b) The acoustic gyrotropic tensor components obey the conditions $d_{ijklm} = d_{jiklm} = d_{ijlkm} = d_{jilkm}$, $d_{ijklm} = -d_{klijm}$. Replace the indices of d_{ijklm} by a reduced-index notation $d_{pq,m}$, where p is the single index corresponding to ij, q is the single index corresponding to kl, and p and q run from 1 to 6 while m runs from 1 to 3. By inspecting the 6×6 matrix of components $d_{pq,m}$ for a fixed value of m, determine the total number of independent components that are possible for the components of the acoustic gyrotropic tensor.

CHAPTER 6

The Geometry of X-Ray Diffraction

6–1 The Ewald Construction

This chapter describes the geometry of some of the experimental arrangements that are used in x-ray diffraction. Basically, these systems facilitate shifts from one Bragg reflection to another, and the geometric relationships involved are expressed concisely by the Ewald sphere construction shown in Fig. 6–1. Point O of Fig. 6–1 is the origin of the reciprocal lattice, and $\mathbf{s}_0$ is a vector of length $1/\lambda$ (where λ is the wavelength) along the direction of the incident beam. A sphere of radius $1/\lambda$ is constructed such that its center lies on the incident beam and its surface passes through O. Point P is a reciprocal lattice point that lies on the surface of the sphere. That is,

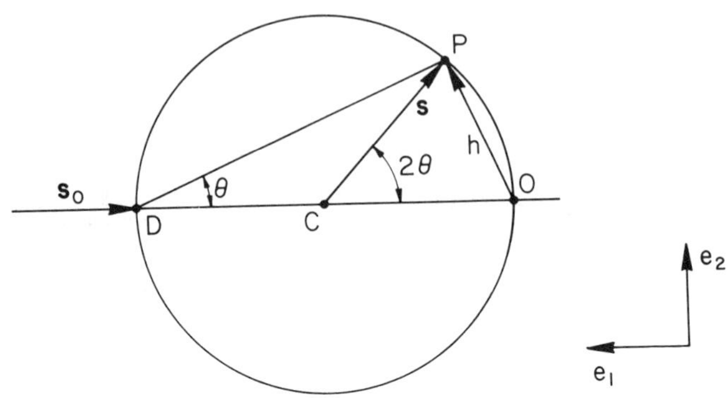

FIGURE 6–1 Ewald sphere construction.

$$OP = \mathbf{h} = h_i \mathbf{a}^i \tag{6–1}$$

The length of **h** is $1/d$, where d is the interplanar spacing. The sine of angle PDO, therefore, is $\lambda/2d$, and this angle may thus be identified with the Bragg angle θ. Figure 6–1 shows that the set of planes characterized by reciprocal lattice vector **h** is in diffracting position when **h** terminates on the sphere. Furthermore, the diffracted (or reflected) beam is in the direction CP, which we shall designate as vector **s** of length $1/\lambda$. The Ewald construction vastly simplifies the complex geometric problems of bringing planes successively into reflecting position and predicting the directions of the reflected rays. From Fig. 6–1,

$$\mathbf{h} = \mathbf{s} - \mathbf{s}_0 \tag{6–2}$$

Calculation of $\mathbf{h} \cdot \mathbf{h}$ leads to Bragg's law:

$$\lambda = 2d \sin \theta \tag{6–3}$$

It is often convenient to denote the location of a reciprocal lattice point by cylindrical coordinates: Z is measured along the rotation axis from a plane perpendicular to the rotation axis through the origin; X is the radial coordinate that gives the distance from the rotation axis; and τ is the angular coordinate that gives the angle between the two planes determined by $\mathbf{s}_0$ and the rotation axis and by **h** and the rotation axis. The angle between the projections of **s** and $\mathbf{s}_0$ onto the plane normal to the rotation axis is denoted Υ. These quantities are depicted in Fig.

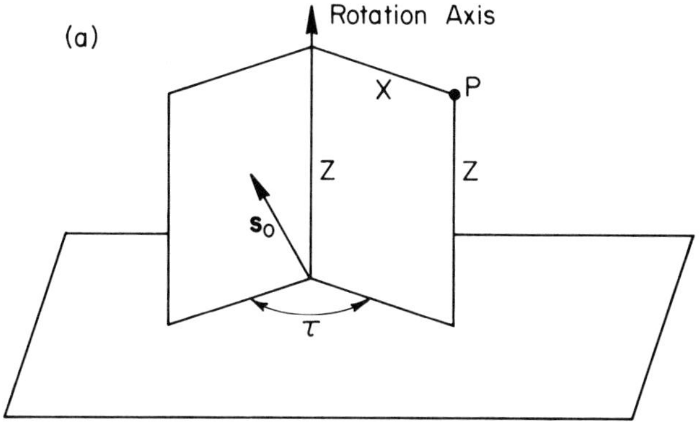

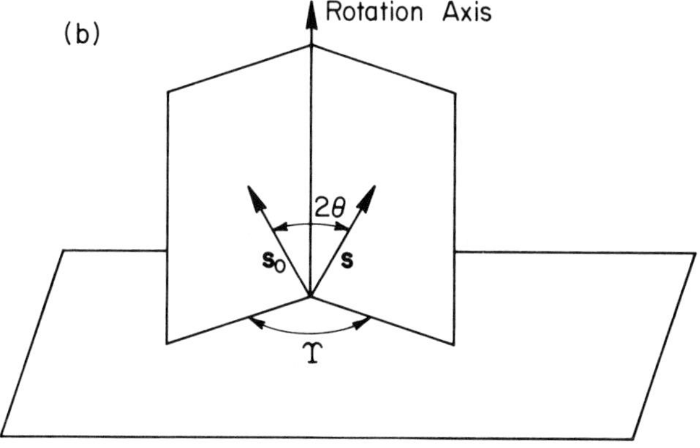

Figure 6–2 (a) Cylindrical coordinates X, Z and τ. (b) Relationships among $\mathbf{s}$, $\mathbf{s}_0$, 2θ, and Υ.

6–2. Much of the crystallographic literature (see especially Buerger, 1942) uses the dimensionless reciprocal lattice coordinates defined by

$$\sigma = \lambda \mathbf{h} \tag{6–4}$$

$$\zeta = \lambda Z \tag{6–5}$$

$$\xi = \lambda X \tag{6–6}$$

$$(\sigma)^2 = (\zeta)^2 + (\xi)^2, \qquad (h)^2 = (Z)^2 + (X)^2 \tag{6–7}$$

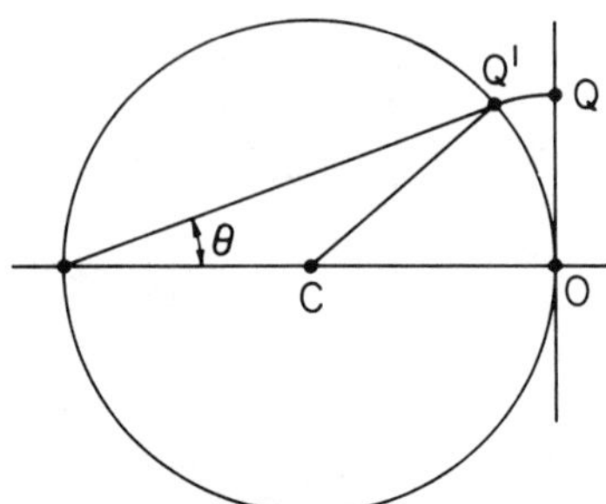

FIGURE 6–3 Rotation of reciprocal lattice point onto the sphere of reflection.

Our choice of h, Z, and X as variables avoids specific dependence of the coordinates upon a particular experiment.

EXERCISE 6–1 Show that the calculation of the length of $\mathbf{s} - \mathbf{s}_0$ leads to Bragg's law.

EXERCISE 6–2 Referring to Fig. 6–3, suppose that a crystal is rotated about an axis normal to the paper through point O. Let Q be a reciprocal lattice point lying on the tangent OQ.

(a) Calculate the angle of rotation in terms of the Bragg angle θ required to bring Q into reflecting position on the surface of the sphere.

(b) Use Fig. 6–3 to deduce the maximum σ that will give Bragg reflection.

6–2 Rotating Crystal Coordinates

For a crystal oriented about a direct lattice vector, say $\mathbf{a}_3$, a reciprocal lattice vector $\mathbf{h}$ may be resolved into components $\mathbf{Z}$ (parallel to $\mathbf{a}_3$) and $\mathbf{X}$ (normal to $\mathbf{a}_3$). The vector $\mathbf{Z}$ is the projection of $\mathbf{h}$ onto $\mathbf{a}_3$, which by Eq. (2–13) is

$$\mathbf{Z} = h_3\mathbf{a}_3/(a_3)^2 \tag{6–8}$$

Hence,

$$\mathbf{X} = \mathbf{h} - \mathbf{Z} = h_i\mathbf{a}^i - h_3\mathbf{a}_3/(a_3)^2 \tag{6–9}$$

Expressing $\mathbf{a}_3$ in terms of the $\mathbf{a}^i$ yields

$$\mathbf{X} = (h_1 - g_{13}h_3/g_{33})\mathbf{a}^1 + (h_2 - g_{23}h_3/g_{33})\mathbf{a}^2 \tag{6–10}$$

The vector $\mathbf{X} - h_1\mathbf{a}^1 - h_2\mathbf{a}^2$ represents the offset from the $\mathbf{a}_3$ axis of the origin of the net at level h_3.

Analysis of the diffraction pattern produced by a crystal rotating in an x-ray beam is facilitated by referring vectors **s** and $\mathbf{s}_0$ to a cartesian coordinate system defined with its origin at the crystal, with $\mathbf{e}_3$ along the rotation axis of the crystal directed away from the spindle of the instrument, and with $\mathbf{e}_1$ along $-\mathbf{s}_0$ as in Fig. 6–1. Letting Υ be the angle between $-\mathbf{e}_1$ and the projection of **s** onto the $\mathbf{e}_1 - \mathbf{e}_2$ plane, and letting ν be the complement of the angle between **s** and the rotation axis, **s** is given by

$$\mathbf{s} = -\frac{\cos\Upsilon\cos\nu}{\lambda}\mathbf{e}_1 + \frac{\sin\Upsilon\cos\nu}{\lambda}\mathbf{e}_2 + \frac{\sin\nu}{\lambda}\mathbf{e}_3 \tag{6-11}$$

These angles are shown in Fig. 6–4. The condition that the angle between **s** and $\mathbf{s}_0$ be 2θ leads to the relationship

$$\cos 2\theta = \cos\Upsilon\cos\nu \tag{6-12}$$

The angle ν may be eliminated from Eq. (6–11) to give

$$\mathbf{s} = -\frac{\cos 2\theta}{\lambda}\mathbf{e}_1 + \frac{\cos 2\theta\tan\Upsilon}{\lambda}\mathbf{e}_2 + \frac{(\cos^2\Upsilon - \cos^2 2\theta)^{1/2}}{\lambda\cos\Upsilon}\mathbf{e}_3 \tag{6-13}$$

The angles in these equations may be expressed in terms of two of the variables *h, X,* and *Z*. By a rearrangement of Bragg's law,

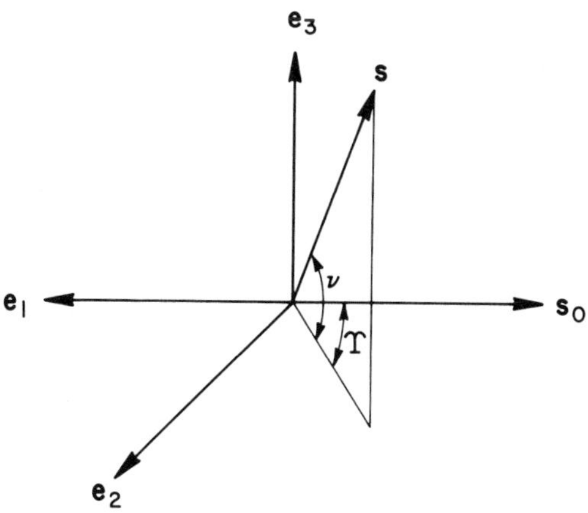

FIGURE 6–4 Angles ν and Υ in rotation of crystal about $\mathbf{e}_3$.

$$\cos 2\theta = 1 - \lambda^2 h^2/2 = 1 - \sigma^2/2 \tag{6-14}$$

and from Eq. (6-11),

$$Z = \sin \nu/\lambda \tag{6-15}$$

Therefore,

$$\cos \Upsilon = \frac{2 - \lambda^2 h^2}{2(1 - \lambda^2 Z^2)^{1/2}} \tag{6-16}$$

and

$$\mathbf{s} = -\frac{2 - \lambda^2 h^2}{2\lambda}\mathbf{e}_1 + \frac{(4\lambda^2 X^2 - \lambda^4 h^4)^{1/2}}{2\lambda}\mathbf{e}_2 + Z\mathbf{e}_3 \tag{6-17}$$

From Eqs. (6-2) and (6-13),

$$\mathbf{h} = -\frac{\cos 2\theta - 1}{\lambda}\mathbf{e}_1 + \frac{\cos 2\theta \tan \Upsilon}{\lambda}\mathbf{e}_2 + \frac{(\cos^2 \Upsilon - \cos^2 2\theta)^{1/2}}{\lambda \cos \Upsilon}\mathbf{e}_3 \tag{6-18}$$

The intensity of a diffracted beam is proportional to the time that the scattering plane spends in reflecting position. This time is inversely proportional to the velocity at which the reciprocal lattice cuts through the sphere of reflection (assuming a constant velocity). In calculating the intensity of an x-ray reflection, therefore, a factor L, called the Lorentz factor, is included; L is inversely proportional to the component normal to the sphere of reflection of the velocity of motion of $\mathbf{h}$. A small rotation $-\delta$ about $\mathbf{e}_3$ converts the vector $\mathbf{h}$ given by Eq. (6-18) into

$$\mathbf{h} = \left(-\frac{\cos 2\theta - 1}{\lambda} + \delta\frac{\cos 2\theta \tan \Upsilon}{\lambda}\right)\mathbf{e}_1 + \left(\delta\frac{\cos 2\theta - 1}{\lambda} + \frac{\cos 2\theta \tan \Upsilon}{\lambda}\right)\mathbf{e}_2 + \frac{(\cos^2\Upsilon - \cos^2 2\theta)^{1/2}}{\lambda \cos \Upsilon}\mathbf{e}_3 \tag{6-19}$$

Differentiating with respect to time and denoting $d\delta/dt$ by ω,

$$\frac{d\mathbf{h}}{dt} = \omega \frac{\cos 2\theta \tan \Upsilon}{\lambda} \mathbf{e}_1 + \omega \frac{\cos 2\theta - 1}{\lambda} \mathbf{e}_2 \qquad (6\text{–}20)$$

By Exercise 1–61, the component of $d\mathbf{h}/dt$ along $\mathbf{s}$ is proportional to $\mathbf{s} \cdot d\mathbf{h}/dt$, and Eqs. (6–13) and (6–20) lead to the angular dependence of L:

$$1/L = \cos 2\theta \tan \Upsilon \qquad (6\text{–}21)$$

which is converted by Eq. (6–12) into

$$1/L = \sin \Upsilon \cos \nu \qquad (6\text{–}22)$$

The intensity calculation includes also a polarization factor, which for initially unpolarized radiation is $(1 + \cos^2 2\theta)/2$; thus,

$$I \propto \frac{1 + \cos^2 2\theta}{\sin \Upsilon \cos \nu} |F|^2 \qquad (6\text{–}23)$$

where F is the structure factor for the reflection.

Although techniques for aligning crystals are rather remote from the main subject matter of this book, it is appropriate to treat the general problem of changing the orientation of a crystal from one vector to another. The usual orientation apparatus consists of a goniometer head with two mutually perpendicular arcs (see Fig. 6–5). A lattice direction of the crystal is aligned along $\mathbf{e}_3$ by adjusting the arcs by rotations about axes in the $\mathbf{e}_1$–$\mathbf{e}_2$ plane. The arc nearer the base of the goniometer head will be labeled B (for bottom), and the smaller arc nearer the crystal will be labeled T (for top). The angle ϕ of rotation about $\mathbf{e}_3$ will be chosen so that when $\phi = 0$ the T axis is along $-\mathbf{e}_1$ and the B axis is along $\mathbf{e}_2$ (at least some instruments are manufactured with this property). A positive ϕ rotation rotates from $\mathbf{e}_1$ to $\mathbf{e}_2$; at $\phi = 90°$ the T axis is along $-\mathbf{e}_2$. Suppose that a crystal oriented initially about unit vector $\mathbf{t}$ is to be reoriented about unit vector $\mathbf{u}$. Complete specification of the problem is obtained by giving the angle ϕ_0 at which some vector $\mathbf{v}$ is along $\mathbf{e}_2$ normal to the x-ray beam: such information is obtained readily from a zero-level Weissenberg photograph (see Section 6–3). The steps in the determination of the new settings are as follows:

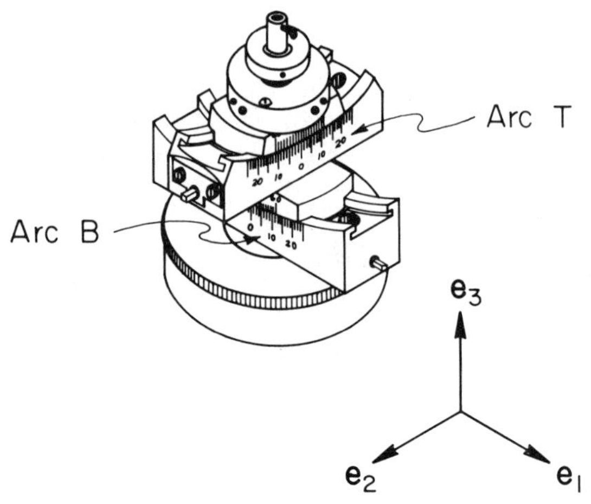

FIGURE 6–5. A goniometer head. The vector triad shows the orientation of the goniometer head when $\phi = 0°$.

1. Calculate $\mathbf{t} \cdot \mathbf{u}$ and find χ, the angle between $\mathbf{t}$ and $\mathbf{u}$. Take $\chi > 0$. (See Fig. 6–6 for these angles.)

2. Evaluate $\mathbf{t} \wedge \mathbf{u}$ in crystal coordinates.

3. Evaluate $\mathbf{v} \cdot (\mathbf{t} \wedge \mathbf{u})$ and determine $\cos \psi$, where ψ is the angle between $\mathbf{t} \wedge \mathbf{u}$ and $\mathbf{v}$.

4. Evaluate $\mathbf{v} \wedge (\mathbf{t} \wedge \mathbf{u})$. This is parallel to $\mathbf{t}$. Letting $\mathbf{v} \wedge (\mathbf{t} \wedge \mathbf{u}) = k\mathbf{t}$, the sign of ψ is the sign of k.

5. Let $\mathbf{t}'$, $\mathbf{u}'$, and $\mathbf{v}'$ be unit vectors in the directions of $\mathbf{t}$, $\mathbf{u}$, and $\mathbf{v}$, respectively. Since $\mathbf{t}' = \mathbf{e}_3$ and $\mathbf{v}' = \mathbf{e}_2$, it follows that

$$\mathbf{u}' = [\sin \chi \cos \psi, \sin \chi \sin \psi, \cos \chi]$$

6. Rotate through $-\phi_0$ about $\mathbf{e}_3$.

$$\mathbf{u}'' = \begin{pmatrix} \cos \phi_0 & \sin \phi_0 & 0 \\ -\sin \phi_0 & \cos \phi_0 & 0 \\ 0 & 0 & 1 \end{pmatrix} \mathbf{u}'$$

7. Set B to 0.

$$\mathbf{u}''' = \begin{pmatrix} \cos B_0 & 0 & -\sin B_0 \\ 0 & 1 & 0 \\ \sin B_0 & 0 & \cos B_0 \end{pmatrix} \mathbf{u}''$$

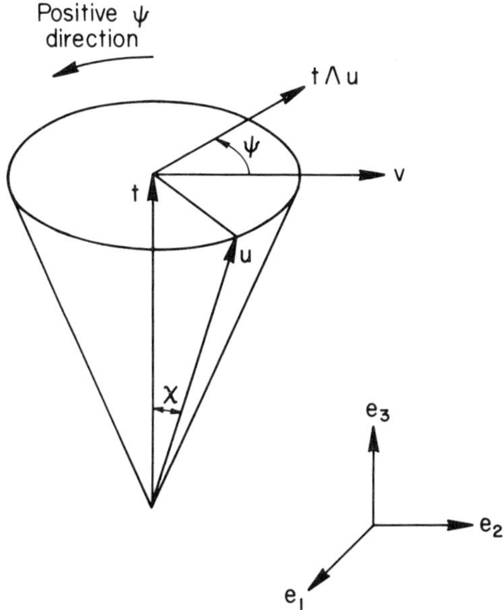

FIGURE 6–6. Alignment angles χ and ψ. The crystal orientation is to be changed from **t** to **u**.

where B_0 is the original value of B.

8. Adjust T to place **u** in the $\mathbf{e}_1$–$\mathbf{e}_3$ plane

$$\mathbf{u}'''' = \begin{pmatrix} 1 & 0 & 0 \\ 0 & \cos \Delta T & -\sin \Delta T \\ 0 & \sin \Delta T & \cos \Delta T \end{pmatrix} \mathbf{u}'''$$

where ΔT, the required change in T, is given by

$$\tan \Delta T = u'''^{2}/u'''^{3}$$

9. Adjust B to align **u** along $\mathbf{e}_3$.

$$\mathbf{u}''''' = \begin{pmatrix} \cos B & 0 & \sin B \\ 0 & 1 & 0 \\ -\sin B & 0 & \cos B \end{pmatrix} \mathbf{u}''''$$

where

$$\tan B = -u''''^{1}/u''''^{3}$$

The net results of these adjustments are

$$\tan \Delta T = \frac{\sin(\psi - \phi_0)}{\sin B_0 \cos(\psi - \phi_0) + \cos B_0 \cot \chi} \quad (6\text{–}24)$$

$$\tan B = \{-\cos B_0 \cos(\psi - \phi_0) + \sin B_0 \cot \chi\} / \{\sin \Delta T \sin(\psi - \phi_0) + \cos \Delta T [\sin B_0 \cos(\psi - \phi_0) + \cos B_0 \cot\chi]\} \quad (6\text{–}25)$$

$$T = T_0 + \Delta T \quad (6\text{–}26)$$

These formulas can be solved for $\psi - \phi_0$ and χ in terms of the goniometer head angles to give

$$\tan(\psi - \phi_0) = \frac{\sin \Delta T}{\sin B_0 \cos \Delta T - \tan B \cos B_0} \quad (6\text{–}27)$$

$$\cot \chi = \sin(\psi - \phi_0)(\sin B_0 \tan B + \cos B_0 \cos \Delta T)/\sin \Delta T \quad (6\text{–}28)$$

EXERCISE 6–3 Substitute values for g_{ij} into Eq. (6–10) to obtain a formula for **X** for a triclinic crystal in terms of the unit cell dimensions.

EXERCISE 6–4 Suppose that a crystal is oriented about $\mathbf{a}_3$, and $\mathbf{a}^1$ is along $\mathbf{s}_0$. Derive a formula for ϕ, the angle of rotation about $\mathbf{a}_3$ necessary to make **X** perpendicular to $\mathbf{s}_0$.

EXERCISE 6–5 Suppose that a rotation photograph is taken with a cylindrical film wrapped about the crystal, with the axis of the cylinder coaxial with **Z**. Starting with Eq. (6–17), derive formulas for coordinates y and z of a reflection on the unwrapped and developed film, where y and z are the $\mathbf{e}_2$ and $\mathbf{e}_3$ components, respectively. The radius of the film is r_F.

EXERCISE 6–6 From Eq. (6–17), find the coordinates y and z of the point at which the reflected beam from a rotating crystal strikes a flat film at distance D from the crystal.

EXERCISE 6–7 A monoclinic crystal for which $a_1 = 10.50$, $a_2 = 8.64$, $a_3 = 6.13$ Å, $\beta = 112.30°$, is oriented about $\mathbf{a}_1$. Weissenberg photographs (see Section 6–3) show that $\mathbf{a}^2$ is normal to the x-ray beam when θ is 28.52°. The goniometer head settings are $T_0 = +14.03°$, $B_0 = -16.57°$. Determine the goniometer head settings that will orient the crystal about [201].

EXERCISE 6–8 When a crystal was aligned about a vector **t**, the goniometer head settings were $T_0 = +12.3°$, $B_0 = -16.8°$. The arcs were changed to orient the crystal about vector **u**, giving new arc settings $T = -14.5°$, $B = +7.2°$. Calculate the angle between **t** and **u**.

6–3 Weissenberg Coordinates

A Weissenberg camera is basically a rotation camera that has been equipped as follows:

1. The camera can be inclined about $\mathbf{e}_2$.
2. The cylindrical film can undergo translation back and forth parallel to the rotation axis $\mathbf{t}$ of the crystal. This translational motion is synchronized with the rotation.
3. A cylindrical metal screen permits the reflections for only one layer (one value of h_3 for an $\mathbf{a}_3$ orientation) to reach the film.

The Weissenberg camera thus selects one layer line at a time from the rotation pattern and spreads it out over the entire film in a manner that permits the positions of each reflection to be correlated with the angle of rotation of the crystal. Essentially, therefore, a Weissenberg photograph is a distorted map of one reciprocal lattice net. The classic treatment of Weissenberg geometry is Buerger's book (1942).

The most advantageous Weissenberg technique uses *equi-inclination* geometry, in which the camera is turned in the $\mathbf{e}_1$–$\mathbf{e}_3$ plane so that for a given layer the incident beam $\mathbf{s}_0$ makes an angle $90° + \nu$ with the axis $\mathbf{t}$ about which the crystal is oriented, and the diffracted beam $\mathbf{s}$ for every reflection in the layer makes angle $90° - \nu$ with $\mathbf{t}$. The advantage of the equi-inclination method is that reflections from points lying on straight lines of the reciprocal lattice nets fall on curves whose shapes are independent of the layer. This geometry is depicted in Fig. 6–7, which shows the plane $\mathbf{e}_1$–$\mathbf{e}_3$. The axes $\mathbf{e}_i$ are assumed to be fixed in space. The cartesian axes $\mathbf{e}_j'$ are inclined to the $\mathbf{e}_i$ by rotation through ν about $\mathbf{e}_2$. The crystal's rotation axis is parallel with $\mathbf{e}_3'$. The diffracted rays from a particular layer form a cone of radiation of half-angle $90° - \nu$. The angle Υ, as defined in Section 6–2, is the angle between $-\mathbf{e}_1'$ and the projection of $\mathbf{s}$ onto the $\mathbf{e}_1'$–$\mathbf{e}_2'$ plane. Corresponding to Eq. (6–11),

$$\mathbf{s} = -\frac{\cos\nu\cos\Upsilon}{\lambda}\mathbf{e}_1' + \frac{\cos\nu\sin\Upsilon}{\lambda}\mathbf{e}_2' + \frac{\sin\nu}{\lambda}\mathbf{e}_3' \qquad (6\text{–}29)$$

and the incident beam vector is

$$\mathbf{s}_0 = -\frac{\cos\nu}{\lambda}\mathbf{e}_1' - \frac{\sin\nu}{\lambda}\mathbf{e}_3' \qquad (6\text{–}30)$$

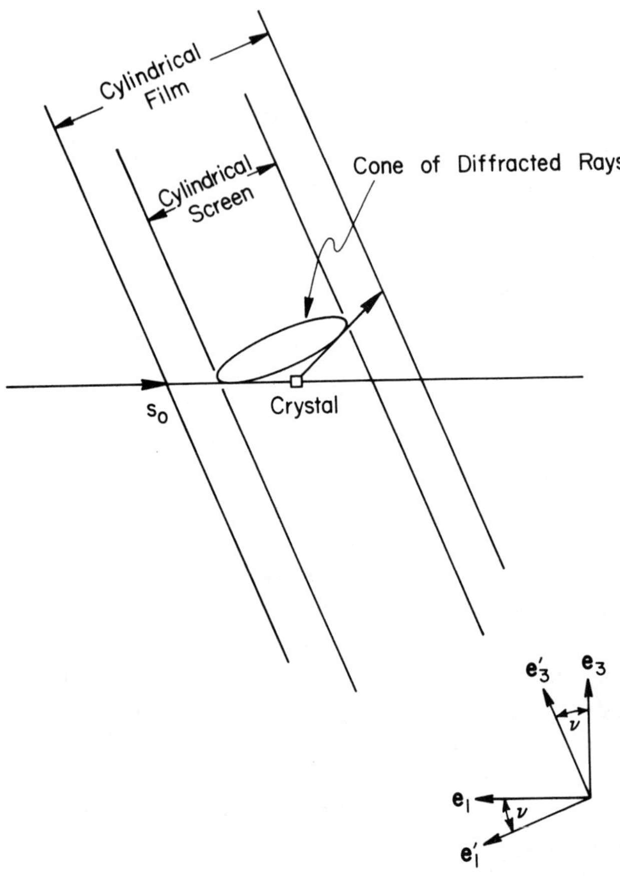

FIGURE 6-7. Equi-inclination Weissenberg geometry.

The requirement that the angle between $\mathbf{s}$ and $\mathbf{s}_0$ be 2θ leads to the relationship [compare Eq. (6-12) for the normal-beam case]

$$\cos \theta = \cos (\Upsilon/2) \cos \nu \tag{6-31}$$

The various angles, of course, can be expressed in terms of any two of the variables h, X, and Z. From $(\mathbf{s} - \mathbf{s}_0) \cdot \mathbf{e}_3'$,

$$\sin \nu = \lambda Z/2 \tag{6-32}$$

and the combination of Eqs. (6–14), (6–31), and (6–32) leads to

$$\cos \Upsilon = \frac{4 - 2\lambda^2 h^2 + \lambda^2 Z^2}{4 - \lambda^2 Z^2} \tag{6–33}$$

The diffracted beam **s** travels through the strategically placed slit in the layer-line screen and strikes the cylindrical film at a distance $r_F \Upsilon$ along the arc of the film. A convenient camera dimension is $r_F = 90/\pi$ mm, in which case the length $r_F \Upsilon$ in millimeters is numerically equal to the value of $\Upsilon/2$ in degrees. Thus, measurement of the distance of a spot from the center of the film gives the value of Υ, and the Bragg angle and hence the value of the interplanar spacing can be determined by means of Eqs. (6–31) and (6–3).

A common mechanism for taking Weissenberg photographs translates the film 1 mm for every two degrees of crystal rotation; thus, a crystal rotation of 180° corresponds to a translation of 90 mm.

To illustrate the use of these equations, suppose that a monoclinic crystal for which $a_1 = 5.00$, $a_2 = 8.00$, $a_3 = 6.00$ Å, $\beta = 105.0°$, is oriented about $\mathbf{a}_3$, and we want to determine where reflections (312) and (022) will appear with radiation of wavelength 1.5418 Å. For (312)

$$(h)^2 = (312)\mathbf{g}^* \begin{pmatrix} 3 \\ 1 \\ 2 \end{pmatrix} = 0.6314$$

and $h = 0.795$. Similarly, for (022), $h = 0.426$. Both of these reflections are in the second layer, for which Eq. (6–8) gives $Z = 2/6.00 = 0.333$, and Eq. (6–32) gives $\nu = 14.89°$. From Eq. (6–33) [not Eq. (6–16), which applies to rotation photographs or normal-beam Weissenberg photographs!], $\Upsilon_{312} = 70.25°$, $\Upsilon_{022} = 24.45°$. Alternatively, $\theta_{312} = 37.78°$, $\theta_{022} = 19.18°$ [by Eqs. (2–37) and (6–3)], and Eq. (6–31) gives the values of Υ. With a film radius of 28.65 mm, the distances of these spots from the center line of the film are 35.13 mm and 12.23 mm, respectively.

If a straight line of slope 2 is drawn through each of these spots on the film, as shown in Fig. 6–8, the distance along $\mathbf{e}_3'$ between these lines is determined by the angle between the two vectors **X** given by Eq. (6–10). We have $\mathbf{X}_{312} = 3.43\mathbf{a}^1 + \mathbf{a}^2$, and $\mathbf{X}_{022} = 0.43\mathbf{a}^1 + 2\mathbf{a}^2$,

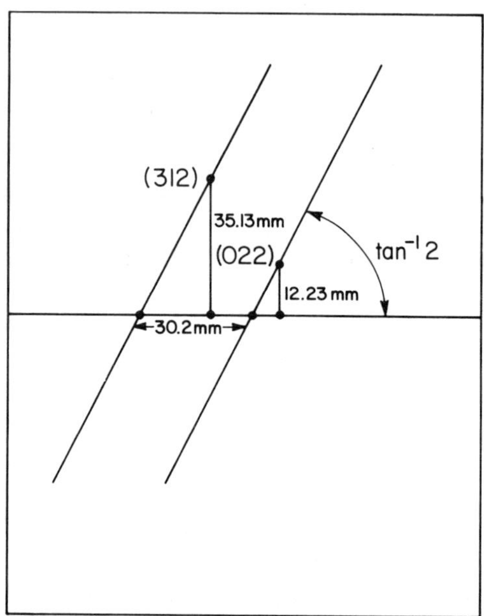

FIGURE 6-8. Location on a Weissenberg film of reflections (312) and (022) for the example of Section 6-3.

and the angle between these vectors is 60.4°. Thus, the film translates 30.2 mm while the crystal rotates from $\mathbf{X}_{312}$ being normal to $\mathbf{e}_1'$ to $\mathbf{X}_{022}$ being normal to $\mathbf{e}_1'$.

EXERCISE 6-9 Show that $X = 2 \cos \nu \sin (\Upsilon/2)/\lambda$.

EXERCISE 6-10 An orthorhombic crystal has $a_1 = 6.00$, $a_2 = 7.00$, $a_3 = 8.00$ Å. Calculate the positions of the reflections (203) and (053) on a Weissenberg film taken with $\lambda = 1.5418$ Å, $r_F = 28.65$ mm, and the crystal oriented about $\mathbf{a}_3$.

EXERCISE 6-11 For the monoclinic crystal used in the example in this section (with $a_1 = 5.00$, $a_2 = 8.00$, $a_3 = 6.00$ Å, $\beta = 105.0°$, oriented about $\mathbf{a}_3$), calculate the positions on a Weissenberg film of the reflections (203), ($\bar{2}$03), and (033). Assume $r_F = 28.65$ mm, $\lambda = 1.5418$ Å.

EXERCISE 6-12 Determine the Lorentz factor for equi-inclination Weissenberg photographs. [Apply the methods of Section 6-2 to Eqs. (6-2), (6-29), and (6-30).]

6–4 Precession Camera

In the precession method (Buerger, 1944, 1964), one axis, which we shall designate $\mathbf{a}_1$, is maintained at a constant angle μ with respect to the incident beam $\mathbf{s}_0$. As $\mathbf{a}_1$ precesses about $\mathbf{s}_0$ the reciprocal lattice rocks through the sphere of reflection. A judiciously placed flat layer screen with an annular slit prevents all reflections except those with a particular value of h_1 from reaching the photographic film. A flat film is used, and the gyrations produced by the mechanical linkage keep both the layer screen and the film parallel with the reciprocal lattice net being photographed. The resulting photograph is an undistorted picture of the reciprocal lattice net.

It is convenient to refer the geometry of the precession camera to two right-handed cartesian coordinate systems centered on the crystal. One of these, with basis vectors $\mathbf{e}_i$, remains fixed in space, with the incident x-ray beam in the $-\mathbf{e}_1$ direction; that is,

$$\mathbf{s}_0 = -\mathbf{e}_1/\lambda \tag{6–34}$$

Axis $\mathbf{e}_3$ of this fixed coordinate system is directed from the crystal along the spindle axis away from the goniometer head when the precession angle μ is set at 0. Axis $\mathbf{e}_2$ is vertical, pointing downward into the paper in Fig. 6–9. These axes are related to the **i**, **j**, **k** used by Waser (1951) in deriving the Lorentz factor for the precession method by

$$\mathbf{e}_1 = \mathbf{j}, \qquad \mathbf{e}_2 = \mathbf{k}, \qquad \mathbf{e}_3 = \mathbf{i} \tag{6–35}$$

This different labeling of the axes maintains consistency with our treatments of the rotation and Weissenberg methods.

With μ still set at 0, the precessing axis, $\mathbf{a}_1$, is aligned in the $\mathbf{e}_1$ direction. Then the precession angle is set by sliding the mechanism through angle μ along the μ arc. The mechanism is now operated by rotating continuously and uniformly about $\mathbf{e}_1$.

The second cartesian coordinate system, with basis vectors $\mathbf{e}_i'$, is coincident with the $\mathbf{e}_i$ system before the μ angle is set. The $\mathbf{e}_i'$ system, however, is fixed to the crystal, so that after setting μ

$$\mathbf{e}_1' = \cos\mu\,\mathbf{e}_1 + \sin\omega\sin\mu\,\mathbf{e}_2 - \cos\omega\sin\mu\,\mathbf{e}_3 \tag{6–36}$$

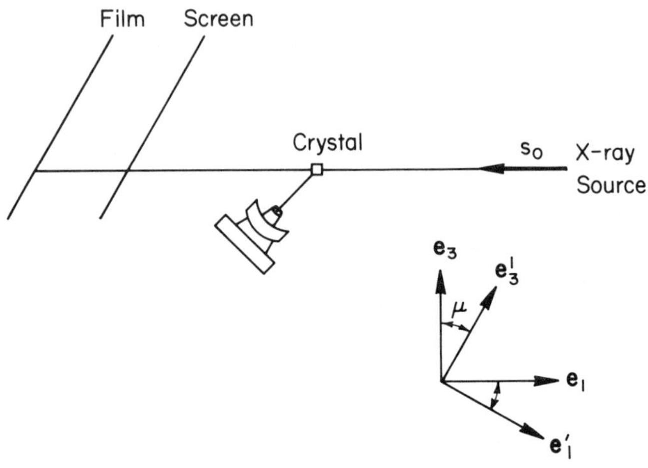

FIGURE 6–9. Precession geometry.

where ω is the angle of rotation of the μ arc about $\mathbf{e}_1$ from its horizontal position. The usual design of the precession camera keeps $\mathbf{e}_3'$, the direction of the spindle axis, horizontal. This constraint, in fact, distorts the motion from being purely precessional.

The complete set of orthonormal $\mathbf{e}_i'$ vectors with the required properties is given by

$$\begin{pmatrix} \mathbf{e}_1' \\ \mathbf{e}_2' \\ \mathbf{e}_3' \end{pmatrix} = \begin{pmatrix} \cos\mu & \sin\omega\sin\mu & -\cos\omega\sin\mu \\ -\dfrac{\sin\omega\sin\mu\cos\mu}{\sin\chi} & \sin\chi & \dfrac{\cos\omega\sin\omega\sin^2\mu}{\sin\chi} \\ \dfrac{\cos\omega\sin\mu}{\sin\chi} & 0 & \dfrac{\cos\mu}{\sin\chi} \end{pmatrix} \begin{pmatrix} \mathbf{e}_1 \\ \mathbf{e}_2 \\ \mathbf{e}_3 \end{pmatrix} \tag{6–37}$$

where

$$\cos\chi = \sin\mu\sin\omega \tag{6–38}$$

In matrix notation, Eq. (6–37) is

$$\mathbf{e}' = \mathbf{G}\mathbf{e} \tag{6–39}$$

Matrix **G** here is an orthogonal matrix with the property

$$\mathbf{G}^{-1} = \overline{\mathbf{G}} \tag{6-40}$$

Hence, the inverse transformation is

$$\mathbf{e} = \overline{\mathbf{G}}\mathbf{e}' \tag{6-41}$$

and the vectors $\mathbf{e}_i$ may be expressed in terms of the $\mathbf{e}_i'$ by using the transpose of the matrix in Eq. (6-37).

The reciprocal lattice vectors are given concisely as

$$\mathbf{X} = X(\sin\tau\,\mathbf{e}_2' + \cos\tau\,\mathbf{e}_3') \tag{6-42}$$

$$\mathbf{Z} = Z\mathbf{e}_1' \tag{6-43}$$

$$\mathbf{h} = Z\mathbf{e}_1' + X(\sin\tau\,\mathbf{e}_2' + \cos\tau\,\mathbf{e}_3') \tag{6-44}$$

where X, Z, and τ are the cylindrical coordinates described in Section 6-1.

In the moving coordinate system, the incident beam vector is

$$\mathbf{s}_0 = \frac{-\sin\chi\cos\mu\,\mathbf{e}_1' + \sin\omega\sin\mu\cos\mu\,\mathbf{e}_2' - \cos\omega\sin\mu\,\mathbf{e}_3'}{\lambda\sin\chi} \tag{6-45}$$

From Eq. (6-2),

$$\mathbf{s} = \frac{[-\cos\mu + \lambda Z]}{\lambda}\,\mathbf{e}_1' + \frac{[\sin\omega\sin\mu\cos\mu + \lambda X\sin\chi\sin\tau]}{\lambda\sin\chi}\,\mathbf{e}_2'$$

$$+ \frac{[-\cos\omega\sin\mu + \lambda X\sin\chi\cos\tau]}{\lambda\sin\chi}\,\mathbf{e}_3' \tag{6-46}$$

Defining ν as the angle between $\mathbf{s}$ and $-\mathbf{e}_1'$, Eq. (6-46) leads to

$$\cos\nu = \cos\mu - \lambda Z = \cos\mu - \zeta \tag{6-47}$$

Letting the projections of $\mathbf{s}$ and $\mathbf{s}_0$ onto the plane normal to $\mathbf{e}_1'$ be, respectively, $\mathbf{R}$ and $\mathbf{R}_0$,

$$\mathbf{R} = \mathbf{s} - (-\cos \mu/\lambda + Z)\mathbf{e}_1' \tag{6–48}$$

$$\mathbf{R}_0 = \mathbf{s}_0 + (\cos \mu/\lambda)\mathbf{e}_1' \tag{6–49}$$

The lengths of $\mathbf{R}$ and $\mathbf{R}_0$ are $\sin \nu/\lambda$ and $\sin \mu/\lambda$, respectively (see Exercise 6–13), and the angle Υ between $\mathbf{R}$ and $\mathbf{R}_0$ is given by

$$\cos \Upsilon = \frac{\cos 2\theta - \cos \mu \cos \nu}{\sin \mu \sin \nu} \tag{6–50}$$

If the angle between $\mathbf{X}$ and $\mathbf{R}_0$ is $\pi - \eta$, taking the scalar product of

$$\mathbf{R} = \mathbf{R}_0 + \mathbf{X} \tag{6–51}$$

with itself leads [with the aid of Eqs. (6–42) and (6–49)] to

$$\cos \eta = \frac{\sin^2 \mu - \sin^2 \nu + \xi^2}{2\xi \sin \mu} \tag{6–52}$$

It should be noted that both the first-quadrant and the fourth-quadrant solutions of this equation are valid values of η.

The vector product $\mathbf{X} \wedge \mathbf{R}_0$ leads also to a formula for η:

$$\sin \eta = \frac{\sin \tau \cos \omega + \cos \tau \sin \omega \cos \mu}{\sin \chi} \tag{6–53}$$

By substituting Eqs. (6–45) and (6–46) into Eqs. (6–48) and (6–49) and taking the scalar product of $\mathbf{R}$ with $\mathbf{R}_0$, we obtain

$$\frac{\lambda h^2 - 2Z \cos \mu}{2X \sin \mu} = \frac{\cos \tau \cos \omega - \sin \tau \sin \omega \cos \mu}{\sin \chi} \tag{6–54}$$

Algebraic manipulation of Eqs. (6–47) and (6–52) leads to

$$\lambda h^2 - 2Z \cos \mu = 2X \cos \eta \sin \mu \tag{6–55}$$

so Eq. (6–54) becomes

$$\cos\eta = \frac{\cos\tau\cos\omega - \sin\tau\sin\omega\cos\mu}{\sin\chi} \tag{6–56}$$

Solving Eqs. (6–53) and (6–56) simultaneously for $\sin\omega/\sin\chi$ and $\cos\omega/\sin\chi$,

$$\frac{\sin\omega}{\sin\chi} = \frac{\sin(\eta - \tau)}{\cos\mu} \tag{6–57}$$

$$\frac{\cos\omega}{\sin\chi} = \cos(\eta - \tau) \tag{6–58}$$

For a given reflection, Eqs. (6–57) and (6–58) give two values of ω, one for the positive value and one for the negative value of η given by Eq. (6–52). One of these values of ω corresponds to the entrance of the reciprocal lattice point into the sphere of reflection; as the motion proceeds, the reciprocal lattice point eventually passes through the sphere again, on its way out, and this position is given by the second value of ω.

Dividing Eq. (6–57) by Eq. (6–58),

$$\tan\omega = \frac{\tan(\eta - \tau)}{\cos\mu} \tag{6–59}$$

With these formulas for ω, Eqs. (6–45) and (6–46) become

$$\mathbf{s}_0 = -\cos\mu\,\frac{\mathbf{e}_1'}{\lambda} + \sin(\eta - \tau)\sin\mu\,\frac{\mathbf{e}_2'}{\lambda} - \cos(\eta - \tau)\sin\mu\,\frac{\mathbf{e}_3'}{\lambda} \tag{6–60}$$

$$\begin{aligned}\mathbf{s} = {} & [-\cos\mu + \lambda Z]\,\frac{\mathbf{e}_1'}{\lambda} + [\sin(\eta - \tau)\sin\mu + \lambda X\sin\tau]\,\frac{\mathbf{e}_2'}{\lambda} \\ & + [-\cos(\eta - \tau)\sin\mu + \lambda X\cos\tau]\,\frac{\mathbf{e}_3'}{\lambda}\end{aligned} \tag{6–61}$$

If the rate of change of ω is constant (typically it is 2π radians per minute), the Lorentz correction factor is

$$\frac{1}{L} = \mathbf{s} \cdot \frac{d\mathbf{h}}{d\omega} \tag{6–62}$$

when **s** and **h** are referred to the fixed cartesian basis vectors $\mathbf{e}_i$. In matrix notation this is

$$\frac{1}{L} = \bar{\mathbf{s}}\mathbf{g}\,\frac{d\mathbf{h}}{d\omega} \tag{6–63}$$

where **g** is the metric tensor (equal to **I** for the cartesian system). Since neither **s** nor **g** depends upon ω, Eq. (6–63) may be written

$$\frac{1}{L} = \frac{d}{d\omega}\,(\bar{\mathbf{s}}\mathbf{g}\mathbf{h}) \tag{6–64}$$

The invariance of scalar products under transformation implies that Eq. (6–64) is valid in any coordinate system. In particular, it may be used in the $\mathbf{e}'$ frame, where only **s** is a function of ω, as well as in the **e** frame, where only **h** depends upon ω. Thus, in the $\mathbf{e}'$ frame

$$\frac{1}{L} = \mathbf{h} \cdot \frac{d\mathbf{s}}{d\omega} \tag{6–65}$$

where **h** is given by Eq. (6–44) and **s** by Eq. (6–46).

Carrying out the calculation of the Lorentz factor in the $\mathbf{e}'$ frame is equivalent to Waser's method (1951) of rotating the sphere of reflection through a stationary reciprocal lattice. The calculation could, however, be carried out just as well in the **e** frame, with a fixed sphere and a moving reciprocal lattice.

From Eqs. (6–46), (6–57), and (6–58),

$$\frac{d\mathbf{s}}{d\omega} = \frac{\tan\mu}{\lambda}\,[\cos^2\mu + \sin^2(\eta - \tau)\sin^2\mu]\,[\cos(\eta - \tau)\mathbf{e}_2' + \sin(\eta - \tau)\mathbf{e}_3'] \tag{6–66}$$

$$\mathbf{h} \cdot \frac{d\mathbf{s}}{d\omega} = \frac{X\sin\mu\cos\mu\sin\eta}{\lambda}\,[1 + \sin^2(\eta - \tau)\tan^2\mu] \tag{6–67}$$

The diffracted intensity is proportional to the reciprocal of $\mathbf{h} \cdot (d\mathbf{s}/d\omega)$. The total intensity observed on a film includes contributions from both values of η. Using the expression for $\sin\eta$ that will be derived in Exercise 6–16, and omitting the constant factor λ, Waser's formula for the Lorentz factor is obtained:

$$\frac{1}{L} = \sin \Upsilon \sin \nu \sin \mu \cos \mu \left[\frac{1}{1 + \sin^2 (\eta - \tau) \tan^2\mu} + \frac{1}{1 + \sin^2(\eta + \tau) \tan^2\mu}\right]^{-1} \qquad (6\text{–}68)$$

This is usually coupled with the polarization factor, which for initially unpolarized radiation is

$$p = \frac{1 + \cos^2 2\theta}{2} \qquad (6\text{–}69)$$

so that the total correction is $1/Lp$.

A numerical example will illustrate the method of calculation. Let $X = 1.000$ Å^{-1}, $Z = 0$, $\tau = 60°$, $\lambda = 0.7107$ Å, $\mu = 30°$. We calculate $\nu = 30°$, $\theta = \sin^{-1}(\lambda h/2) = 20.81°$, $\Upsilon = 90.58°$, $\eta = 44.71°$, $\mathbf{h} = 0.8660\mathbf{e}_2' + 0.5000\mathbf{e}_3'$. Although the values of ω are not needed for the calculation, they will help in understanding the motion: Eq. (6–59) gives the ω values $-17.52°$ and $257.19°$, corresponding to $\eta = 44.71°$ and $\eta = -44.71°$, respectively [whereas Eq. (6–59) is useful in calculating the values of ω, Eqs. (6–57) and (6–58) are helpful in determining the proper quadrants]. The incident and diffracted beam vectors for the case $\eta = 44.71°$, $\omega = -17.52°$ are

$$\mathbf{s}_0 = -1.2186\mathbf{e}_1' - 0.1856\mathbf{e}_2' - 0.6786\mathbf{e}_3'$$

$$\mathbf{s} = -1.2186\mathbf{e}_1' + 0.6805\mathbf{e}_2' - 0.1786\mathbf{e}_3'$$

From Eqs. (6–66) and (6–67)

$$\frac{d\mathbf{s}}{d\omega} = 0.6013\mathbf{e}_2' - 0.1644\mathbf{e}_3'$$

$$\mathbf{h} \cdot \frac{d\mathbf{s}}{d\omega} = 0.4386$$

The corresponding results for $\eta = -44.71°$, $\omega = 257.19°$, are

$$\mathbf{s}_0 = -1.2186\mathbf{e}_1' - 0.6805\mathbf{e}_2' + 0.1786\mathbf{e}_3'$$

$$\mathbf{s} = -1.2186\mathbf{e}_1' + 0.1856\mathbf{e}_2' + 0.6786\mathbf{e}_3'$$

$$\frac{d\mathbf{s}}{d\omega} = -0.2029\mathbf{e}_2' - 0.7731\mathbf{e}_3'$$

$$\mathbf{h} \cdot \frac{d\mathbf{s}}{d\omega} = -0.5623$$

$$L = \tfrac{1}{2}\,(1/0.4386 + 1/0.5623) = 1/0.4928$$

For this reflection, $p = 0.7793$, so $1/Lp = 0.6323$. The maximum value of $1/Lp$ on the $Z = 0$ level with $\mu = 30°$ occurs at $\tau = 90°$, $\xi = 0.81483$ (see Exercise 6–17); our formulas for $1/Lp$ give 0.68393 at this point (see Exercise 6–18). It is customary to normalize $1/Lp$ to a value of 1.00000 at $\zeta = 0$, $\xi = 0.81483$, $\tau = 90°$, $\mu = 30°$. On this scale, the relative value of $1/Lp$ at the point in our example is $0.6323/0.68393 = 0.9245$. Tables of such normalized values are available (Buerger, 1964; *International Tables for X-Ray Crystallography,* Vol. II).

Another feature of precession photographs that can be elucidated by these vector methods is the doubling of spots that occurs on upper levels if the film position is set incorrectly. If the film for a zero-level photograph is at distance F from the crystal, the vector to the center of the film when $\mu = 0$ is

$$\mathbf{c} = F\lambda\mathbf{s}_0 \tag{6–70}$$

The horizontal bisector of the film lies in the $\mathbf{e}_1$–$\mathbf{e}_3$ plane and is parallel to $\mathbf{e}_3'$. The diffracted beam strikes the film for a zero level when its component along $\mathbf{e}_1'$ is $-F \cos \mu$. By Eq. (6–61) (with $Z = 0$) this results if the diffracted beam vector is

$$\mathbf{v} = F\lambda\mathbf{s} \tag{6–71}$$

For an upper-level photograph the component of $\mathbf{s}$ along $\mathbf{e}_1'$ is $Z - \cos \mu/\lambda$, and the Z in this term will produce a different magnification for each layer unless an appropriate shift of the film is made. The camera is constructed so that the film can be translated along the $\mathbf{e}_1'$ direction. If the film is translated back toward the crystal by distance D, the center of the film is now (Fig. 6-10)

$$\mathbf{c} = F\lambda\mathbf{s}_0 + D\mathbf{e}_1' = -F\mathbf{e}_1 + D\mathbf{e}_1' \qquad (6\text{-}72)$$

Expressing $\mathbf{e}_1$ in terms of the $\mathbf{e}_i'$ [by Eqs. (6-37) and (6-41)] and eliminating ω and χ [by Eqs. (6-57) and (6-58)],

$$\mathbf{c} = (-F\cos\mu + D)\mathbf{e}_1' + F\sin(\eta - \tau)\sin\mu\,\mathbf{e}_2'$$

$$- F\cos(\eta - \tau)\sin\mu\,\mathbf{e}_3' \qquad (6\text{-}73)$$

Vector $\mathbf{v}$, from the crystal to the film in the direction of $\mathbf{s}$, may be scaled to make its $\mathbf{e}_1'$ component also equal to $-F\cos\mu + D$ by multiplying Eq. (6-61) by $(-F\cos\mu + D)/(Z - \cos\mu/\lambda)$; that is,

$$\mathbf{v} = [(-F\cos\mu + D)/(Z - \cos\mu/\lambda)]\mathbf{s} \qquad (6\text{-}74)$$

The position of the spot on the film is

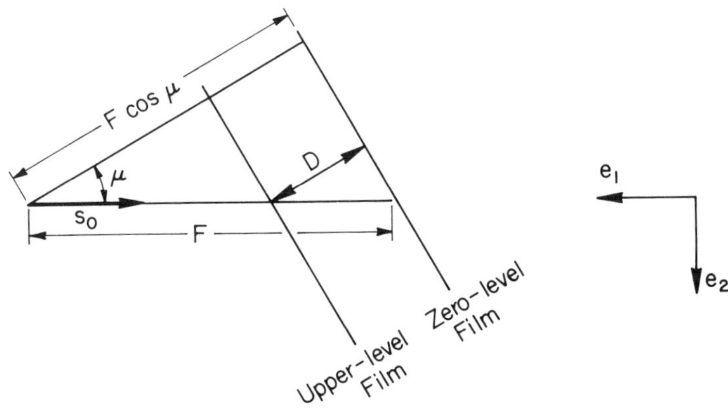

FIGURE 6-10. Film displacement for upper-level precession photograph.

$$\mathbf{v} - \mathbf{c} = \left\{ \frac{[-F\cos\mu + D][\sin(\eta - \tau)\sin\mu + \lambda X\sin\tau]}{\lambda Z - \cos\mu} - F\sin(\eta - \tau)\sin\mu \right\} \mathbf{e}_2' + \left\{ \frac{[-F\cos\mu + D][-\cos(\eta - \tau)\sin\mu + \lambda X\cos\tau]}{\lambda Z - \cos\mu} + F\cos(\eta - \tau)\sin\mu \right\} \mathbf{e}_3' \quad (6\text{–}75)$$

The η (and hence ω) dependence in this equation can be eliminated by setting

$$-F\cos\mu + D = F(\lambda Z - \cos\mu) \quad (6\text{–}76)$$

Therefore, the distance that the film must be advanced is

$$D = F\lambda Z = F\zeta \quad (6\text{–}77)$$

With this setting,

$$\mathbf{v} - \mathbf{c} = F\lambda X(\sin\tau\, \mathbf{e}_2' + \cos\tau\, \mathbf{e}_3') \quad (6\text{–}78)$$

and the scale or magnification of the photograph is the same for every layer.

If the film is advanced by an amount different from D, $\mathbf{v} - \mathbf{c}$ remains dependent upon η and ω, and the diffracted beams for the two values of ω do not intersect the film at the same location; that is, the spot on the film is split. Suppose that an error is made in setting D and the film is advanced by $F\lambda Z + \delta$ instead of $F\lambda Z$. The change in $\mathbf{v} - \mathbf{c}$ is worked out from Eq. (6–75) to be

$$\Delta = \frac{\delta}{\lambda Z - \cos\mu} \{[\sin(\eta - \tau)\sin\mu + \lambda X\sin\tau]\mathbf{e}_2' + [-\cos(\eta - \tau)\sin\mu + \lambda X\cos\tau]\mathbf{e}_3'\} \quad (6\text{–}79)$$

EXERCISE 6–13 Calculate the lengths of vectors $\mathbf{R}$ and $\mathbf{R}_0$ of Eqs. (6–48) and (6–49).

EXERCISE 6–14 As the precession mechanism rotates through angle ω about $\mathbf{e}_1$, the crystal turns about $\mathbf{e}_3'$, the spindle axis, by angle $\Delta\phi$.

(a) Show that $\Delta\phi$ is given by

$$\sin \Delta\phi = \cos \chi = \sin \mu \sin \omega$$

(b) What is the range of oscillation of ϕ as ω covers the range from 0 to 2π?

EXERCISE 6–15 Show that

$$\cos \Upsilon = (\sin^2 \mu + \sin^2 \nu - \xi^2)/(2 \sin \mu \sin \nu)$$

EXERCISE 6–16 Starting with Eq. (6–52), show that

$$\sin \eta = \pm \sin \nu \sin \Upsilon/\xi.$$

EXERCISE 6–17 Write $1/Lp$ as a function of ξ for the case $\mu = 30°$, $Z = 0$, $\tau = 90°$, and find the value of ξ that will maximize $1/Lp$.

EXERCISE 6–18 Calculate the two values of ω, the corresponding vectors $\mathbf{s}_0$, $\mathbf{s}$, and $d\mathbf{s}/d\omega$, and the value of $1/Lp$ at the point $\tau = 90°$, $\xi = 0.81483$, $Z = 0$, $\mu = 30°$, $\lambda = 0.7107$ Å.

EXERCISE 6–19 Calculate $1/Lp$ for the point $X = 0.9849$, $Z = 0.2814$, $\tau = 40°$, $\mu = 30°$, $\lambda = 0.7107$ Å.

EXERCISE 6–20 A precession photograph is taken with $\mu = 30°$ and $F = 60$ mm of a layer with $\zeta = \lambda Z = 0.1000$. The film is erroneously moved forward only 5.50 mm instead of the proper 6.00 mm. Calculate the two positions of the spot given by $\xi = \lambda X = 0.4000$, $\tau = 60°$, and compare with the position at which the combined spot should appear.

6–5 Eulerian Angles

The orientation of a crystal in space may be specified in terms of three Eulerian angles, ϕ, χ, ω, which show the relationships between a set of cartesian axes $\mathbf{e}_i$ that are fixed in space and a set of cartesian axes $\mathbf{e}_i'$ that are attached to the crystal. The fixed axes $\mathbf{e}_i$ will be chosen here as a right-handed set with $\mathbf{e}_1$ along $-\mathbf{s}_0$, $\mathbf{e}_2$ horizontal, and $\mathbf{e}_3$ vertical. The $\mathbf{e}'$ axes are assumed to be initially coincident with the $\mathbf{e}$ axes. The conventions followed here in selecting and specifying the Eulerian angles are those of Hamilton (1974); however, the lack of consistency in the literature and in instrumental design makes it imperative that close attention be paid to the definitions of the angles and the directions of the rotations used in a particular experimental arrangement.

New orientations of the $\mathbf{e}_i'$ axes with respect to the fixed $\mathbf{e}_i$ are achieved by rotating first through ϕ about $\mathbf{e}_3$, then through χ about $\mathbf{e}_1$, and finally through ω about $\mathbf{e}_3$. Following Hamilton, the positive sense of the χ rotation is counterclockwise when viewed from the x-ray source with ω equal to 0 (see Fig. 6–11). The positive sense of the ϕ motion is counterclockwise in Fig. 6–11 when viewed from the top of the circle with $\chi = 0$. Likewise, the positive direction of the ω rotation is counterclockwise. Essentially, the positive direction of each rotation is that which takes $\mathbf{e}_1$ into $\mathbf{e}_2$, $\mathbf{e}_2$ into $\mathbf{e}_3$, and $\mathbf{e}_3$ into $\mathbf{e}_1$.

The effect of the rotations ϕ, χ, and ω on the components of a vector may be written

$$v'^{i} = R^{i}_{j}\, v^{j} \tag{6–80}$$

where the R^{i}_{j} are the elements of the matrix

$$\mathbf{R} = \mathbf{R}(\omega)\mathbf{R}(\chi)\mathbf{R}(\phi) \tag{6–81}$$

in which

$$\mathbf{R}(\phi) = \begin{pmatrix} \cos\phi & -\sin\phi & 0 \\ \sin\phi & \cos\phi & 0 \\ 0 & 0 & 1 \end{pmatrix} \tag{6–82}$$

$$\mathbf{R}(\chi) = \begin{pmatrix} 1 & 0 & 0 \\ 0 & \cos\chi & -\sin\chi \\ 0 & \sin\chi & \cos\chi \end{pmatrix} \tag{6–83}$$

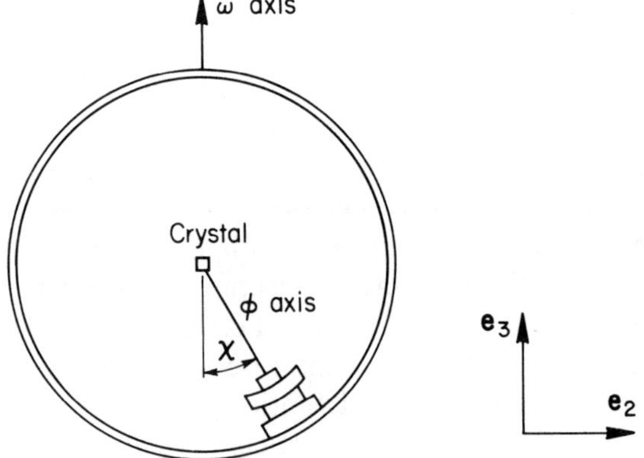

FIGURE 6–11. View of χ circle from x-ray source in Eulerian geometry, with $\omega = 0°$, $\mathbf{e}_1$ outward normal to the page.

$$\mathbf{R}(\omega) = \begin{pmatrix} \cos\omega & -\sin\omega & 0 \\ \sin\omega & \cos\omega & 0 \\ 0 & 0 & 1 \end{pmatrix} \tag{6–84}$$

All three of these matrices are orthogonal, so

$$\mathbf{R}^{-1} = \bar{\mathbf{R}} \tag{6–85}$$

A set of cartesian axes $\mathbf{e}_i'$ initially coincident with the $\mathbf{e}_i$ are rotated to orientations given by the matrix equation

$$\mathbf{e}' = \bar{\mathbf{R}}(\phi)\bar{\mathbf{R}}(\chi)\bar{\mathbf{R}}(\omega)\mathbf{e} \tag{6–86}$$

When written out fully,

$$\mathbf{e}_1' = (\cos\omega\cos\phi - \sin\omega\cos\chi\sin\phi)\mathbf{e}_1 + (\sin\omega\cos\phi + \cos\omega\cos\chi\sin\phi)\mathbf{e}_2 + \sin\chi\sin\phi\,\mathbf{e}_3 \tag{6–87}$$

$$\mathbf{e}_2' = -(\cos\omega\sin\phi + \sin\omega\cos\chi\cos\phi)\mathbf{e}_1 + (-\sin\omega\sin\phi + \cos\omega\cos\chi\cos\phi)\mathbf{e}_2 + \sin\chi\cos\phi\,\mathbf{e}_3 \tag{6–88}$$

$$\mathbf{e}_3' = \sin\omega\sin\chi\,\mathbf{e}_1 - \cos\omega\sin\chi\,\mathbf{e}_2 + \cos\chi\,\mathbf{e}_3 \tag{6–89}$$

The inverse transformation, given by the transpose of matrix **R**, is

$$\mathbf{e}_1 = (\cos\omega\cos\phi - \sin\omega\cos\chi\sin\phi)\,\mathbf{e}_1' + (-\cos\omega\sin\phi - \sin\omega\cos\chi\cos\phi)\,\mathbf{e}_2' + \sin\omega\sin\chi\,\mathbf{e}_3' \tag{6–90}$$

$$\mathbf{e}_2 = (\sin\omega\cos\phi + \cos\omega\cos\chi\sin\phi)\,\mathbf{e}_1' + (-\sin\omega\sin\phi + \cos\omega\cos\chi\cos\phi)\,\mathbf{e}_2' - \cos\omega\sin\chi\,\mathbf{e}_3' \tag{6–91}$$

$$\mathbf{e}_3 = \sin\chi\sin\phi\,\mathbf{e}_1' + \sin\chi\cos\phi\,\mathbf{e}_2' + \cos\chi\,\mathbf{e}_3' \tag{6–92}$$

The principal crystallographic application of Eulerian geometry is in devices that will place reciprocal lattice points on the Ewald sphere of reflection. The usual mechanism aligns reciprocal lattice vector **h** along $\mathbf{e}_2$. Rotation through θ about $\mathbf{e}_3$ then moves the reciprocal lattice points along this row successively into diffracting position. In synchronization with this motion, the detector moves through angle 2θ so that it is positioned properly to receive the diffracted radiation. This positive θ motion is clockwise in Fig. 6–12, opposite in direction to the ω and ϕ rotations.

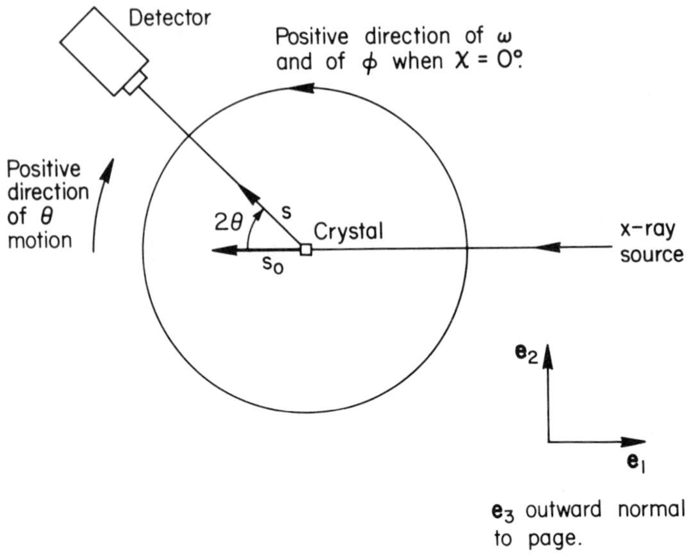

FIGURE 6-12. Positive directions of rotations in Eulerian geometry.

Only two rotations are needed to orient a vector in space (corresponding to the two angles of spherical coordinates), so the ω rotation is redundant. Some of the early instruments, such as the single-crystal orienter developed by T. Furnas and D. Harker, permitted very little ω adjustment. However, the ability to vary ω permits alternative settings of the other two angles in order to avoid instrumental obstacles, and it allows rotation of the crystal about $\mathbf{h}$ (by a combination of ω, ϕ, and χ motions); such motion, for example, may provide data for making empirical absorption corrections. Therefore, most modern single-crystal diffractometers utilizing the Eulerian angle geometry are four-circle instruments (the fourth angle being θ). A method for orienting a reciprocal lattice vector $\mathbf{h}$ along $\mathbf{e}_2$ is to set ω and χ equal to zero, adjust ϕ to place $\mathbf{h}$ in the $\mathbf{e}_2$–$\mathbf{e}_3$ plane, and finally change χ to make $\mathbf{h}$ horizontal. The instrumental settings for this alignment with $\omega = 0°$ will be denoted ϕ_0 and χ_0. An ω rotation now changes $\mathbf{h}$ from $h\mathbf{e}_2$ to

$$\mathbf{h} = h(-\sin\omega\,\mathbf{e}_1 + \cos\omega\,\mathbf{e}_2) \tag{6-93}$$

The vector **h** may be restored to alignment along $\mathbf{e}_2$ by a combination of ϕ and χ adjustments that obey the matrix equation

$$\mathbf{R}(\chi)\mathbf{R}(\phi - \phi_0)\mathbf{R}(-\chi_0) \begin{pmatrix} 0 \\ 1 \\ 0 \end{pmatrix} = \begin{pmatrix} \sin\omega \\ \cos\omega \\ 0 \end{pmatrix} \qquad (6\text{–}94)$$

Effectively, Eq. (6–94) is a rotation of **h** through $-\omega$ about $\mathbf{e}_3$. Equating corresponding components in Eq. (6–94) leads to

$$\sin(\phi - \phi_0)\cos\chi_0 = -\sin\omega \qquad (6\text{–}95)$$

$$\cos\chi\cos(\phi - \phi_0)\cos\chi_0 + \sin\chi\sin\chi_0 = \cos\omega \qquad (6\text{–}96)$$

$$\sin\chi\cos(\phi - \phi_0)\cos\chi_0 - \cos\chi\sin\chi_0 = 0 \qquad (6\text{–}97)$$

Solving both Eq. (6–96) and Eq. (6–97) for $\cos(\phi - \phi_0)$ and equating

$$\sin\chi = \sin\chi_0/\cos\omega \qquad (6\text{–}98)$$

Eqs. (6–95) and (6–97) give

$$\sin(\phi - \phi_0) = -\sin\omega/\cos\chi_0 \qquad (6\text{–}99)$$

$$\cos(\phi - \phi_0) = \cot\chi\tan\chi_0 \qquad (6\text{–}100)$$

Equation 6–98 gives two values of χ, related by $\chi_1 + \chi_2 = 180°$. For each of these, Eqs. (6–99) and (6–100) provide a value of $\phi - \phi_0$. These values of ω, χ, and ϕ provide a means of rotating the reciprocal lattice about **h**; the angle ψ by which the reciprocal lattice is rotated about **h** is positive when measured clockwise as viewed *from* the origin. Defining $\mathbf{R}(\psi)$ as the matrix that gives the new coordinates in the **e** coordinate system,

$$\mathbf{R}(\psi) = \begin{pmatrix} \cos\psi & 0 & \sin\psi \\ 0 & 1 & 0 \\ -\sin\psi & 0 & \cos\psi \end{pmatrix} \qquad (6\text{–}101)$$

The angle ψ is determined from the condition

$$\mathbf{R}(\omega)\mathbf{R}(\chi)\mathbf{R}(\phi - \phi_0)\mathbf{R}(-\chi_0) = \mathbf{R}(\psi) \qquad (6\text{–}102)$$

Equating corresponding elements

$$\cos \psi = \cos \chi / \cos \chi_0 \tag{6-103}$$

$$\sin \psi = \tan \omega \tan \chi_0 \tag{6-104}$$

$$\sin \Delta\phi = -\sin \psi / \sin \chi \tag{6-105}$$

While any direction in a crystal may serve as the orientation axis, some applications find it advantageous to use either a reciprocal lattice vector or a direct lattice vector. It may be easier to align a crystal about a reciprocal lattice vector (by using the reflections at $\chi = 90°$); with such an orientation all reciprocal lattice points along a row parallel to the rotation axis will have the same value of ϕ. On the other hand, alignment about a direct lattice vector will make an entire net accessible to measurement at $\chi = 0°$. Modern computer-controlled diffractometers make such special orientations unnecessary, and it is in fact preferable to align the crystal about a general vector to avoid such problems as double reflection. Our pedagogical goals will be served, however, by examining the particular cases of orientations about reciprocal lattice or direct lattice vectors.

Considering first the case of orientation about a reciprocal lattice vector, say $\mathbf{a}^3$, suppose that the instrumental settings are $\phi(1)$, $\chi(1)$, $\omega(1)$ when $\mathbf{h}(1)$ is along $\mathbf{e}_2$ in reflecting position. A second reciprocal lattice vector $\mathbf{h}(2)$ may be aligned along $\mathbf{e}_2$ by performing the rotations implied by the matrix product

$$\mathbf{R}(\omega(2))\mathbf{R}(\chi(2))\mathbf{R}(\Delta\phi)\mathbf{R}(-\chi(1))\mathbf{R}(-\omega(1))$$

If $\omega(2)$ is taken to be 0, $\Delta\phi$ is the angle between the projections of $\mathbf{h}(1)$ and $\mathbf{h}(2)$ onto the plane normal to $\mathbf{a}^3$. By Eq. (2–45), the projection of vector $\mathbf{h}(I)$ onto the plane normal to $\mathbf{a}^3$ is

$$\mathbf{P}[\mathbf{h}(I)||\mathbf{a}^3] = \mathbf{h}(I) - [\mathbf{h}(I) \cdot \mathbf{a}^3]\mathbf{a}^3/g^{33} \tag{6-106}$$

which may be written

$$\mathbf{P}[\mathbf{h}(I)||\mathbf{a}^3] = h_1(I)\mathbf{a}^1 + h_2(I)\mathbf{a}^2 - [h_1(I)g^{13} + h_2(I)g^{23}]\mathbf{a}^3/g^{33} \tag{6-107}$$

The scalar product of $\mathbf{P}[\mathbf{h}(1)||\mathbf{a}^3]$ and $\mathbf{P}[\mathbf{h}(2)||\mathbf{a}^3]$ provides a value of $\cos \Delta\phi$. The vector product of these two vectors is

$$\mathbf{P}[\mathbf{h}(1)\,||\,\mathbf{a}^3] \wedge \mathbf{P}[\mathbf{h}(2)\,||\,\mathbf{a}^3] = [h_1(1)h_2(2) - h_2(1)h_1(2)]\mathbf{a}^3/(g^{33}V) \tag{6–108}$$

The sign of $\Delta\phi$ is therefore the sign of $h_1(1)h_2(2) - h_2(1)h_1(2)$, and $\Delta\phi$ is independent of $h_3(\mathrm{I})$.

Angle $\chi(I)$ is the complement of the angle between $\mathbf{h}(I)$ and $\mathbf{a}^3$; hence,

$$\sin\chi(I) = [h_i(I)\mathbf{a}^i]\cdot\mathbf{a}^3/(h(I)a^3) = h_i(I)g^{i3}/(h(I)a^3) \tag{6–109}$$

For values of $\omega(2)$ other than 0, new values of ϕ and χ must be calculated from Eqs. (6–98) to (6–100).

The method of calculation will be illustrated by the following example. Consider a monoclinic crystal for which $a_1 = 7.680$, $a_2 = 9.040$, $a_3 = 8.510$ Å, $\beta = 104.30°$. The crystal is oriented with $\mathbf{a}^3$ parallel to $\mathbf{e}_3$, and the (210) reflection is observed at $\omega = 0°$, $\phi = 122.63°$. It is desired to calculate χ for (210) and ϕ and χ for ($\bar{1}32$) for the case $\omega = 0°$. The lengths of the vectors $\mathbf{h}(210)$ and $\mathbf{h}(\bar{1}32)$ are, respectively, 0.2906 and 0.4134. The projections of these vectors onto the plane normal to $\mathbf{a}^3$ are (2, 1, -0.5474) and ($\bar{1}$, 3, 0.2737), with respective lengths 0.2829 and 0.3565. The scalar product of these two projected vectors with each other is 0.002802, from which $\Delta\phi$ is 88.41°. Since $h_1(1)h_2(2) - h_2(1)h_1(2)$ is positive (equal to 7) and the coordinate system presumably is right-handed, ϕ for ($\bar{1}32$) is $122.63 + 88.41 = 211.04°$. The scalar product (210) • (001) is equal to 0.008050, so the angle between these vectors is 76.80°, and χ for (210) is $90 - 76.80 = 13.20°$. Similarly, χ for ($\bar{1}32$) is 30.42°. Suppose now that it is desired to measure ($\bar{1}32$) at $\omega = 30.00°$. By Eq. (6–98), $\sin\chi = \sin 30.42°/\cos 30.00°$, and χ is 35.78° or 144.22°. With $\chi = 35.78°$, Eqs. (6–99) and (6–100) give $\Delta\phi = -35.44°$, so ϕ equals $211.04 - 35.44 = 175.60°$. With $\chi = 144.22°$, $\Delta\phi = -144.56°$, $\phi = 211.04 - 144.56 = 66.48°$.

These results find compact expression via an orientation matrix $\mathbf{U}$ which relates the crystal's reciprocal axes to the cartesian vectors $\mathbf{e}_i'$.

$$\mathbf{a}^i = U^{ij}\mathbf{e}_j' \tag{6–110}$$

We align $\mathbf{h}$ along $\mathbf{e}_2$:

$$\mathbf{h} = h_i\mathbf{a}^i = h\mathbf{e}_2 \tag{6–111}$$

Inserting Eq. (6–110) and taking the scalar product with $\mathbf{e}'_k$,

$$h_i U^{ik} = h\mathbf{e}_2 \cdot \mathbf{e}'_k \qquad (6\text{–}112)$$

or

$$\bar{\mathbf{h}}\mathbf{U} = \bar{\mathbf{Y}} \qquad (6\text{–}113)$$

Equations (6–87), (6–88), and (6–89) give the components of vector **Y** in the **e**′ frame:

$$\bar{\mathbf{Y}} = h(\cos\phi \sin\omega + \sin\phi \cos\chi \cos\omega, \ -\sin\phi \sin\omega + \cos\phi \cos\chi \cos\omega, \ -\sin\chi \cos\omega) \qquad (6\text{–}114)$$

Matrix **U** may be determined from

$$\mathbf{U} = \mathbf{H}^{-1}\mathbf{T} \qquad (6\text{–}115)$$

where $\mathbf{H}^{-1}$ is the inverse of the matrix

$$\mathbf{H} = \begin{pmatrix} h_1(1) & h_2(1) & h_3(1) \\ h_1(2) & h_2(2) & h_3(2) \\ h_1(3) & h_2(3) & h_3(3) \end{pmatrix} \qquad (6\text{–}116)$$

and $h_i(j)$ is the ith index of reflection j. Row i of matrix **T** consists of the components in the **e**′ frame of vector $\bar{\mathbf{Y}}$ for reflection i; these components are given by Eq. (6–114); that is,

$$T_{ij} = Y_j(i) = h(i)\,\mathbf{e}_2 \cdot \mathbf{e}'_j \qquad (6\text{–}117)$$

The scalar products in Eq. (6–117) may be computed from Eq. (6–91) with results that are of course consistent with Eq. (6–114). If ϕ, ω, and χ are known for three reflections, the indices of reflection i form row i of **H**, and the components of $\bar{\mathbf{Y}}$ for reflection i form row i of **T**.

If only two reflections have been observed, the parameters of a third reciprocal lattice vector to be used in determining **U** may be obtained from

$$\mathbf{h}(3) = \mathbf{h}(1) \wedge \mathbf{h}(2) \qquad (6\text{–}118)$$

where the vector product is taken in crystal coordinates, and

$$\mathbf{Y}(3) = \mathbf{Y}(1) \wedge \mathbf{Y}(2) \tag{6–119}$$

where the vector product is between vectors in the cartesian $\mathbf{e}'$ frame given by Eq. (6–114). If more observations are made than are needed to determine **U**, a least-squares refinement of the elements of **U** may be carried out. Once **U** has been found, the angular settings of other reflections may be computed from Eq. (6–113). By Eq. (6–110),

$$\mathbf{g}^* = \mathbf{U}\overline{\mathbf{U}} \tag{6–120}$$

so the determination of **U** leads to knowledge of the metric tensor.

The use of the orientation matrix will be illustrated by application to the example we have already treated in this section. With an $\mathbf{a}^3$ orientation, one of the rows of **H** is $\overline{\mathbf{h}}(1) = (001)$, for which χ is 90°, and $\overline{\mathbf{Y}}(1)$ is worked out from Eq. (6–114) to be (0, 0, -0.1213). For the (210) reflection, $\overline{\mathbf{h}}(2) = (210)$, $\omega = 0°$, $\chi = 13.20°$, $\phi = 122.63°$, $\overline{\mathbf{Y}}(2) = 0.2906(0.8199, -0.5250, -0.2284)$. Then $\mathbf{h}(1) \wedge \mathbf{h}(2) = V^*(2\mathbf{a}_2 - \mathbf{a}_1)$, which becomes

$$\mathbf{h}(1) \wedge \mathbf{h}(2) = V^*[2(g_{21}\mathbf{a}^1 + g_{22}\mathbf{a}^2 + g_{23}\mathbf{a}^3) - (g_{11}\mathbf{a}^1 + g_{12}\mathbf{a}^2 + g_{13}\mathbf{a}^3)]$$

The covariant crystal components of $\mathbf{h}(3)$ are $(-0.1030, 0.2855, 0.0282)$. Equation (6–119) gives $\mathbf{Y}(3) = -0.01850\mathbf{e}_1' - 0.02886\mathbf{e}_2'$. Then

$$\mathbf{H} = \begin{pmatrix} 0 & 0 & 1 \\ 2 & 1 & 0 \\ -0.1030 & 0.2855 & 0.0282 \end{pmatrix}$$

$$\mathbf{T} = \begin{pmatrix} 0 & 0 & -0.1213 \\ 0.2383 & -0.1526 & -0.0664 \\ -0.01850 & -0.02886 & 0 \end{pmatrix}$$

and Eq. (6–115) produces

$$\mathbf{U} = \mathbf{H}^{-1}\mathbf{T} = \begin{pmatrix} 0.1284 & -0.0218 & -0.0332 \\ -0.0185 & -0.1090 & 0 \\ 0 & 0 & -0.1213 \end{pmatrix}$$

For the $(\bar{1}32)$ reflection, $(-1, 3, 2)\mathbf{U} = (-0.1838, -0.3050, -0.2094)$, $h = 0.4134$, and by comparing with Eq. (6–114) with $\omega = 0°$, $\sin\phi\cos\chi = -0.1838/0.4134$, $\cos\phi\cos\chi = -0.3050/0.4134$, $\sin\chi = -(-0.2094)/0.4134$. The angles are $\chi = 30.42°$, $\phi = 211.07°$.

If a crystal is oriented about a direct lattice vector, say $\mathbf{a}_3$, the projection of reciprocal lattice vector $(h_1h_2h_3)$ onto the plane normal to $\mathbf{a}_3$ is $(h_1'\ h_2'\ 0)$, where $h_i' = h_i - h_3g_{3i}/g_{33}$. Then $\Delta\phi$ is the angle between $h_1'(1)\mathbf{a}^1 + h_2'(1)\mathbf{a}^2$ and $h_1'(2)\mathbf{a}^1 + h_2'(2)\mathbf{a}^2$. The vector product of these two vectors is

$$\mathbf{P}[\mathbf{h}(1)\,||\,\mathbf{a}_3]\ \Lambda\ \mathbf{P}[\mathbf{h}(2)\,||\,\mathbf{a}_3] = [h_1'(1)h_2'(2) - h_2'(1)h_1'(2)]\mathbf{a}_3/V \tag{6–121}$$

so the sign of $\Delta\phi$ is the sign of $h_1'(1)h_2'(2) - h_2'(1)h_1'(2)$. The χ setting is given by

$$\sin\chi = h_3/(ha_3) \tag{6–122}$$

These results are readily generalized to orientation about any vector.

EXERCISE 6–21 For the monoclinic crystal in the example in this section, calculate ϕ and χ for the case $\omega = 0°$, $\mathbf{a}^3$ orientation, for the reflections (200), (211), (132), (032), $(0\bar{3}\bar{2})$, (130), (030), and $(\bar{1}30)$.

EXERCISE 6–22 An orthorhombic crystal for which $a_1 = 7.80$, $a_2 = 11.22$, $a_3 = 12.05$ Å was oriented about $\mathbf{a}^2$. The (200) reflection was observed with $\omega = 0°$, $\phi = 260.42°$.

(a) For the case $\omega = 0°$, calculate ϕ and χ for (i) (022), (ii) (432), (iii) $(\bar{4}32)$.

(b) For the case $\omega = 45°$, calculate ϕ and χ for the following reflections. Explain any cases for which solution is impossible. (i) (022), (ii) (432), (iii) $(\bar{4}32)$, (iv) (200).

EXERCISE 6–23 The axes of the crystal of Exercise 6–22 may be transformed to a nonprimitive unit cell with $a_1 = 19.71$, $a_2 = 25.33$, $a_3 = 13.66$ Å, $\alpha = 79.88°$, $\beta = 116.85°$, $\gamma = 70.25°$. With the crystal oriented about the new $\mathbf{a}^3$, the reflection with new indices $(4\bar{2}\bar{2})$ was found at $\omega = 0°$, $\phi = 260.42°$. Using the dimensions of this oblique cell, calculate ϕ and χ for $(\bar{2}\bar{4}2)$, $(6\bar{8}\bar{1})$, and $(\bar{1}\bar{0}, 0, 7)$, where the indices are all for the new cell and ω is maintained at 0°.

EXERCISE 6–24 A hexagonal crystal for which $a_1 = 9.76$, $a_3 = 12.52$ Å is aligned about $\mathbf{a}^1$. The (250) reflection is observed at $\phi = 75.38°$ with $\omega = 0°$.

(a) Calculate χ for (250).

(b) Calculate ϕ and χ for (503).
(c) Calculate ϕ and χ for (523).
(d) Calculate χ for (200).

EXERCISE 6-25 A certain reflection is observed with instrumental settings ω, χ, ϕ, 2θ.

(a) Show that this reflection may be observed also at the settings. (i) $(\pi + \omega, -\chi, \pi + \phi, 2\theta)$; (ii) $(-\omega, \pi - \chi, \pi + \phi, 2\theta)$; (iii) $(\pi - \omega, \pi + \chi, \phi, 2\theta)$.

(b) Show that the reflection may be observed also at the settings (i) $(-\omega, \pi + \chi, \phi, -2\theta)$; (ii) $(\pi - \omega, \pi - \chi, \pi + \phi, -2\theta)$; (iii) $(\omega, -\chi, \pi + \phi, -2\theta)$; (iv) $(\pi + \omega, \chi, \phi, -2\theta)$.

EXERCISE 6-26 A reflection is aligned with $\chi_0 = 40°$. What is the maximum value of ω that can be used for observation of this reflection? Calculate χ and $\Delta\phi$ for this reflection with this maximum ω.

EXERCISE 6-27 A reflection is aligned with $\chi_0 = 90°$. At what values of ω may this reflection be observed?

EXERCISE 6-28 A reciprocal lattice vector **h** is aligned with $\omega_0 = 0°$, $\chi_0 = 60°$. If ω is changed to 20°, what values of $\Delta\phi$ and χ will restore **h** to alignment?

6-6 Kappa Geometry

Kappa geometry replaces the χ motion of Eulerian geometry by a kappa rotation about an axis that is at angle α with respect to $\mathbf{e}_3$ (the cartesian vectors will be defined as in Section 6-5). Kappa geometry is unique to the Enraf-Nonius CAD-4 diffractometer, the design of which is protected by law. The angle of α is set at 50° in the CAD-4 diffractometer, and the cartesian coordinates of the kappa axis are $[-\sin 50°, 0, \cos 50°]$ when $\omega = 0°$. The positive direction of the kappa rotation will be taken as counterclockwise when viewed from above. The ϕ and ω rotations will also be positive in a counterclockwise direction, as in the Eulerian geometry discussion. Kappa geometry is diagrammed in Fig. 6-13.

The change of coordinates produced by rotation of a vector through angle $\varkappa$ about the kappa axis is equivalent to

$$\mathbf{R}(\varkappa) = \begin{pmatrix} \cos\alpha & 0 & -\sin\alpha \\ 0 & 1 & 0 \\ \sin\alpha & 0 & \cos\alpha \end{pmatrix} \begin{pmatrix} \cos\varkappa & -\sin\varkappa & 0 \\ \sin\varkappa & \cos\varkappa & 0 \\ 0 & 0 & 1 \end{pmatrix} \begin{pmatrix} \cos\alpha & 0 & \sin\alpha \\ 0 & 1 & 0 \\ -\sin\alpha & 0 & \cos\alpha \end{pmatrix} =$$

$$\begin{pmatrix} \cos\varkappa\cos^2\alpha + \sin^2\alpha & -\sin\varkappa\cos\alpha & -\cos\alpha\sin\alpha\,(1 - \cos\varkappa) \\ \sin\varkappa\cos\alpha & \cos\varkappa & \sin\varkappa\sin\alpha \\ -\cos\alpha\sin\alpha\,(1 - \cos\varkappa) & -\sin\varkappa\sin\alpha & \cos\varkappa\sin^2\alpha + \cos^2\alpha \end{pmatrix} \tag{6–123}$$

The combined effect of ω, $\varkappa$, and ϕ motions is represented by the matrix $\mathbf{R}$ given by

$$\mathbf{R} = \mathbf{R}(\omega, \varkappa, \phi) = \mathbf{R}(\omega)\mathbf{R}(\varkappa)\mathbf{R}(\phi) \tag{6–124}$$

where $\mathbf{R}(\omega)$ is given by Eq. (6–84), $\mathbf{R}(\varkappa)$ by Eq. (6–123), and $\mathbf{R}(\phi)$ by Eq. (6–82). In terms of $\mathbf{R}$, a vector $\mathbf{v}$ that has components v^i when ω, $\varkappa$, and ϕ are all set to 0° will have components

$$v'^j = R^j{}_i\, v^i \tag{6–125}$$

for general values of the angles. A set of cartesian axes initially coincident with the $\mathbf{e}_i$ will become

$$\begin{aligned} \mathbf{e}_1' = {} & [-\cos\alpha\sin\varkappa\sin(\omega + \phi) + \cos\varkappa\cos(\omega + \phi) + \\ & \quad \sin^2\alpha\cos\omega\cos\phi\,(1 - \cos\varkappa)]\mathbf{e}_1 \\ & + [\cos\alpha\sin\varkappa\cos(\omega + \phi) + \cos\varkappa\sin(\omega + \phi) + \\ & \quad \sin^2\alpha\sin\omega\cos\phi\,(1 - \cos\varkappa)]\mathbf{e}_2 \\ & + [-\cos\alpha\sin\alpha\cos\phi\,(1 - \cos\varkappa) - \sin\alpha\sin\phi\sin\varkappa]\mathbf{e}_3 \end{aligned} \tag{6–126}$$

$$\begin{aligned} \mathbf{e}_2' = {} & [-\cos\alpha\sin\varkappa\cos(\omega + \phi) - \cos\varkappa\sin(\omega + \phi) - \\ & \quad \sin^2\alpha\cos\omega\sin\phi\,(1 - \cos\varkappa)]\mathbf{e}_1 \\ & + [-\cos\alpha\sin\varkappa\sin(\omega + \phi) + \cos\varkappa\cos(\omega + \phi) - \\ & \quad \sin^2\alpha\sin\omega\sin\phi\,(1 - \cos\varkappa)]\mathbf{e}_2 \\ & + [\cos\alpha\sin\alpha\sin\phi\,(1 - \cos\varkappa) - \sin\alpha\cos\phi\sin\varkappa]\mathbf{e}_3 \end{aligned} \tag{6–127}$$

$$\begin{aligned} \mathbf{e}_3' = {} & [-\cos\alpha\sin\alpha\cos\omega\,(1 - \cos\varkappa) - \sin\alpha\sin\omega\sin\varkappa]\mathbf{e}_1 \\ & + [-\cos\alpha\sin\alpha\sin\omega\,(1 - \cos\varkappa) + \sin\alpha\cos\omega\sin\varkappa]\mathbf{e}_2 \\ & + [\sin^2\alpha\cos\varkappa + \cos^2\alpha]\mathbf{e}_3 \end{aligned} \tag{6–128}$$

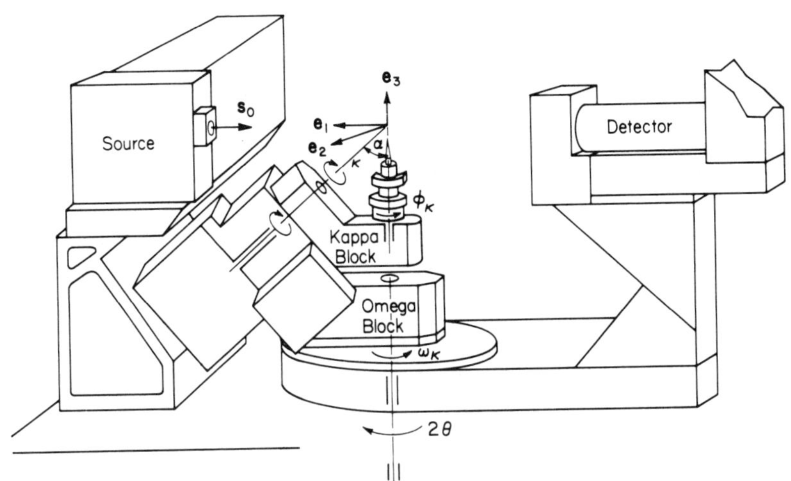

FIGURE 6–13. Kappa geometry. Adapted from operating manual for ENRAF-NONIUS CAD 4 diffractometer (angles ω, ϕ, and $\varkappa$ are opposite in sign to those of Enraf-Nonius). (By permission of ENRAF-NONIUS Service Corp., Bohemia, New York.)

Comparison with Eqs. (6–87), (6–88), and (6–89) reveals the relationships for equivalent orientations between the Eulerian angles ω_E, χ, and ϕ_E and the kappa angles $\omega_\varkappa$, $\varkappa$, and $\phi_\varkappa$. The ratios of the $\mathbf{e}_1$ and $\mathbf{e}_2$ components of $\mathbf{e}_3'$ lead to

$$\omega_E = \omega_\varkappa + \delta \tag{6–129}$$

where

$$\delta = \tan^{-1}(\cos\alpha \tan\varkappa/2) \tag{6–130}$$

Similarly, the $\mathbf{e}_3$ components of $\mathbf{e}_1'$ and $\mathbf{e}_2'$ lead to

$$\phi_E = \phi_\varkappa + \delta \tag{6–131}$$

Equating the $\mathbf{e}_3$ components of $\mathbf{e}_3'$ produces

$$\cos\chi = \cos\varkappa \sin^2\alpha + \cos^2\alpha \tag{6–132}$$

Complete specification of χ is obtained by combining any one of the components involving sin χ in Eq. (6–87), (6–88), or 6–89) with Eq. (6–129), (6–130), or (6–131) to give

$$\sin \chi = -\sin \alpha \sin \kappa/\cos \delta \tag{6–133}$$

For cases where cos δ vanishes, a convenient variation of this is

$$\sin \chi = \frac{-\sin 2\alpha \sin^2 \kappa/2}{\sin \delta} \tag{6–134}$$

It must be emphasized that the angles ω_E, ω_κ, ϕ_E, ϕ_κ, and κ (all except χ) are opposite in sign to those described in the operation manual of the Enraf–Nonius CAD 4 diffractometer (1977). As defined here, the sign of χ is opposite to that of δ and κ. With this sign convention, Eq. (6–132) may be reduced to

$$\sin \chi/2 = -\sin \alpha \sin \kappa/2 \tag{6–135}$$

These equations permit calculation of the Eulerian angles from the kappa angles. For the reverse calculation, δ must be obtained in terms of χ. From Eqs. (6–133) and (6–135),

$$\cos \delta = \frac{\cos (\kappa/2)}{\cos (\chi/2)} \tag{6–136}$$

and, using Eq. (6–132),

$$\cos \delta = \frac{[\cos^2\chi/2 - \cos^2\alpha]^{1/2}}{\cos (\chi/2) \sin \alpha} \tag{6–137}$$

From Eqs. (6–134) and (6–135),

$$\sin \delta = -\cot \alpha \tan (\chi/2) \tag{6–138}$$

A reciprocal lattice vector **h** may be placed in diffracting position by adjusting κ to place the vector in the equatorial plane, then changing ω to align it along $\mathbf{e}_2$. If a vector has been aligned with instrumental settings ϕ_o, κ_o, ω_o, a new set of angles, ϕ, κ, and ω, corresponding to rotation about **h** must obey

$$\mathbf{R}(\psi)\begin{pmatrix}0\\1\\0\end{pmatrix} = \begin{pmatrix}0\\1\\0\end{pmatrix} \tag{6–139}$$

where

$$\mathbf{R}(\psi) = \mathbf{R}(\omega)\mathbf{R}(\varkappa)\mathbf{R}(\Delta\phi)\overline{\mathbf{R}}(\varkappa_o)\overline{\mathbf{R}}(\omega_o) \tag{6–140}$$

in which

$$\phi = \phi_o + \Delta\phi \tag{6–141}$$

and $\mathbf{R}(\omega)$ is given by Eq. (6–84), $\mathbf{R}(\varkappa)$ by Eq. (6–123), and $\mathbf{R}(\Delta\phi)$ by Eq. (6–82). The new $\varkappa$ is determined by requiring the vanishing of the component along $\mathbf{e}_3$ of the vector

$$\mathbf{R}(\varkappa)\mathbf{R}(\Delta\phi)\overline{\mathbf{R}}(\varkappa_o)\overline{\mathbf{R}}(\omega_o)\begin{pmatrix}0\\1\\0\end{pmatrix}$$

The new ω then returns the vector to the $\mathbf{e}_2$ direction. Explicit expressions for $\varkappa$, ω, and ψ will not be presented; the calculations are best carried out in matrix form.

An orientation matrix $\mathbf{U}$ for kappa geometry may be determined by procedures parallel to those applied to Eulerian geometry. The main difference is that in this case

$$\begin{aligned}\overline{\mathbf{Y}} = h[&\cos\alpha\sin\varkappa\cos(\omega+\phi) + \cos\varkappa\sin(\omega+\phi) + \\ &\sin^2\alpha\sin\omega\cos\phi(1-\cos\varkappa), -\cos\alpha\sin\varkappa\sin(\omega+\phi) + \\ &\cos\varkappa\cos(\omega+\phi) - \sin^2\alpha\sin\omega\sin\phi(1-\cos\varkappa), \\ &-\cos\alpha\sin\alpha\sin\omega(1-\cos\varkappa) + \sin\alpha\cos\omega\sin\varkappa]\end{aligned} \tag{6–142}$$

Three such row vectors $\overline{\mathbf{Y}}(i)$ constitute matrix $\mathbf{T}$ according to Eq. (6–117), $\mathbf{H}$ is defined by Eq. (6–116), and $\mathbf{U}$ is obtained from Eq. (6–115). The complexity of Eq. (6–142), however, suggests that it is easier to solve first for the Eulerian angles by means of the equations of Section 6–5, then to convert to kappa angles using Eqs. (6–137), (6–138), (6–132), (6–133), (6–129), and (6–131).

EXERCISE 6–29 If $\varkappa = 70°$, $\phi_\varkappa = 160°$, $\omega_\varkappa = -25°$, and $\alpha = 50$;

(a) Calculate the components of $\mathbf{e}_i'$ from Eqs. (6–126), (6–127), and (6–128).
(b) Calculate χ, ϕ_E, and ω_E.
(c) Calculate the components of $\mathbf{e}_i'$ from Eqs. (6–87), (6–88), and (6–89).

EXERCISE 6–30 A reciprocal lattice vector $\mathbf{h}$ is aligned along $\mathbf{e}_2$ with instrumental settings $\phi_o = 130.00°$, $\varkappa_o = 25.62°$, $\omega_o = 18.04°$. If ϕ is changed to 140.00°, what will be the new $\varkappa$ and ω settings required to maintain $\mathbf{h}$ in diffracting position? What is the angle of rotation ψ?

EXERCISE 6–31 Calculate the kappa geometry angles ($\phi_\varkappa$, $\omega_\varkappa$, $\varkappa$) for the (210) and ($\bar{1}32$) reflections of the crystal discussed in the example of Section 6–5.

CHAPTER 7

Curvilinear Coordinates

7-1 Introduction

In a general curvilinear coordinate system, the lengths and directions of the basis vectors depend upon the coordinates; hence, the elements of the metric tensors also are functions of the coordinates. A justification for including a treatment of curvilinear coordinates in a book on rectilinear geometry is that it will complete an introduction to tensor analysis and remove barriers to further study. A more immediate purpose is that an understanding of the principles of curvilinear geometry will assist in clarifying many aspects of the special case of rectilinear systems. Finally, some problems involving functions of rectilinear coordinates are rendered tractable by the general methods of tensor analysis.

7-2 Spherical Coordinates

The development of curvilinear geometry is introduced most easily by means of an example. The most widely used curvilinear coordinate system is the spherical coordinate system, in which the location of a point is specified by the coordinates r, θ, and ϕ, as shown in Fig. 7-1. Point P is at distance r from the origin O. Vector **OP** makes angle θ with the cartesian axis $\mathbf{e}_3$, and the projection of **OP** onto the $\mathbf{e}_1$–$\mathbf{e}_2$ plane makes angle ϕ with $\mathbf{e}_1$.

The relationships between the cartesian coordinates and the spherical coordinates are given by

$$x = r \sin\theta \cos\phi \tag{7-1}$$

$$y = r \sin\theta \sin\phi \tag{7-2}$$

$$z = r \cos\theta \tag{7-3}$$

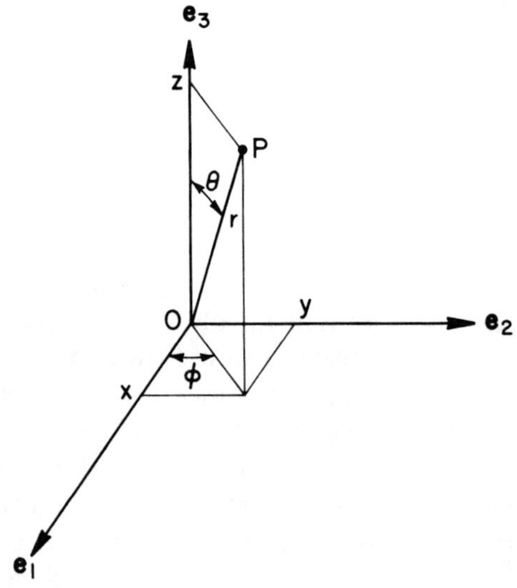

FIGURE 7-1. Spherical coordinates.

where for convenience the cartesian coordinates are denoted by x, y, z instead of x^1, x^2, x^3. For differential changes in the coordinates,

$$dx = \frac{\partial x}{\partial r}dr + \frac{\partial x}{\partial \theta}d\theta + \frac{\partial x}{\partial \phi}d\phi \tag{7-4}$$

$$dy = \frac{\partial y}{\partial r}dr + \frac{\partial y}{\partial \theta}d\theta + \frac{\partial y}{\partial \phi}d\phi \tag{7-5}$$

$$dz = \frac{\partial z}{\partial r}dr + \frac{\partial z}{\partial \theta}d\theta + \frac{\partial z}{\partial \phi}d\phi \tag{7-6}$$

In matrix notation,

$$\begin{pmatrix} dx \\ dy \\ dz \end{pmatrix} = \begin{pmatrix} \frac{\partial x}{\partial r} & \frac{\partial x}{\partial \theta} & \frac{\partial x}{\partial \phi} \\ \frac{\partial y}{\partial r} & \frac{\partial y}{\partial \theta} & \frac{\partial y}{\partial \phi} \\ \frac{\partial z}{\partial r} & \frac{\partial z}{\partial \theta} & \frac{\partial z}{\partial \phi} \end{pmatrix} \begin{pmatrix} dr \\ d\theta \\ d\phi \end{pmatrix} \tag{7-7}$$

which is the differential analogue of Eq. (3–12). Thus, if **F** is the matrix that transforms coordinates *from* the cartesian system *to* the curvilinear system,

$$\begin{pmatrix} dx \\ dy \\ dz \end{pmatrix} = \mathbf{F}^{-1} \begin{pmatrix} dr \\ d\theta \\ d\phi \end{pmatrix} \tag{7-8}$$

The elements of matrix $\mathbf{F}^{-1}$ are the partial derivatives in Eq. (7–7). Similarly, the elements of **F** are the partial derivatives of the spherical coordinates with respect to the cartesian coordinates [cf. Eq. (3–49)].

The determinant of $\mathbf{F}^{-1}$ is called the Jacobian of the transformation

$$J = |\mathbf{F}^{-1}| = \frac{\partial(x, y, z)}{\partial(r, \theta, \phi)} \tag{7-9}$$

For the spherical coordinate system

$$\mathbf{F}^{-1} = \begin{pmatrix} \sin\theta\cos\phi & r\cos\theta\cos\phi & -r\sin\theta\sin\phi \\ \sin\theta\sin\phi & r\cos\theta\sin\phi & r\sin\theta\cos\phi \\ \cos\theta & -r\sin\theta & 0 \end{pmatrix} \tag{7–10}$$

$$J = r^2 \sin\theta \tag{7–11}$$

Equation (3–43) expressed that fact that upon transformation the volume of the unit cell defined by the basis vectors is multiplied by the determinant of the transformation matrix. A similar result applies here, although it is a differential volume element that is involved. Thus,

$$dV = dx\,dy\,dz = r^2 \sin\theta\, dr\, d\theta\, d\phi \tag{7–12}$$

By analogy with the results of Chapter 3, if $\mathbf{F}$ is the transformation matrix for contravariant coordinates, the matrix $\mathbf{G} = \overline{\mathbf{F}}^{-1}$ is the transformation matrix for covariant basis vectors. If $\mathbf{a}_i$ are the basis vectors associated with the spherical system, the transpose of the inverse of $\mathbf{F}^{-1}$ produces

$$\begin{pmatrix} \mathbf{e}_1 \\ \mathbf{e}_2 \\ \mathbf{e}_3 \end{pmatrix} = \begin{pmatrix} \sin\theta\cos\phi & \dfrac{\cos\theta\cos\phi}{r} & -\dfrac{\sin\phi}{r\sin\theta} \\ \sin\theta\sin\phi & \dfrac{\cos\theta\sin\phi}{r} & \dfrac{\cos\phi}{r\sin\theta} \\ \cos\theta & -\dfrac{\sin\theta}{r} & 0 \end{pmatrix} \begin{pmatrix} \mathbf{a}_1 \\ \mathbf{a}_2 \\ \mathbf{a}_3 \end{pmatrix} \tag{7–13}$$

The inverse of this transformation is

$$\begin{pmatrix} \mathbf{a}_1 \\ \mathbf{a}_2 \\ \mathbf{a}_3 \end{pmatrix} = \begin{pmatrix} \sin\theta\cos\phi & \sin\theta\sin\phi & \cos\theta \\ r\cos\theta\cos\phi & r\cos\theta\sin\phi & -r\sin\theta \\ -r\sin\theta\sin\phi & r\sin\theta\cos\phi & 0 \end{pmatrix} \begin{pmatrix} \mathbf{e}_1 \\ \mathbf{e}_2 \\ \mathbf{e}_3 \end{pmatrix} \tag{7–14}$$

These transformations may be written succinctly

$$\mathbf{e} = \overline{\mathbf{F}}\mathbf{a} \tag{7–15}$$

$$\mathbf{a} = \mathbf{G}\mathbf{e} \tag{7–16}$$

By Eq. (3–44), the metric tensor for spherical coordinates is

$$\mathbf{g} = \begin{pmatrix} 1 & 0 & 0 \\ 0 & r^2 & 0 \\ 0 & 0 & r^2 \sin^2 \theta \end{pmatrix} = \overline{\mathbf{F}}^{-1}\mathbf{F}^{-1} = \mathbf{G}\overline{\mathbf{G}} \tag{7–17}$$

The vanishing of the off-diagonal terms of $\mathbf{g}$ establishes that the basis vectors for the spherical system are orthogonal. They are not orthonormal, however: $a_1 = 1$, $a_2 = r$, $a_3 = r \sin \theta$. It is instructive to examine the rotations and stretchings of these basis vectors on moving around in the space. At a point on the $\mathbf{e}_1$ axis, where $\theta = 90°$, $\phi = 0°$, $\mathbf{a}_1$ coincides with $\mathbf{e}_1$ (in direction and in length), $\mathbf{a}_2$ has length r and is in the $-\mathbf{e}_3$ direction, and $\mathbf{a}_3$ has length r and is in the $\mathbf{e}_2$ direction. At a point on the $\mathbf{e}_2$ axis, where $\theta = 90°$ and $\phi = 90°$, $\mathbf{a}_1$ is a unit vector along $\mathbf{e}_2$, $\mathbf{a}_2$ has length r and is in the $-\mathbf{e}_3$ direction, and $\mathbf{a}_3$ has length r and is directed along $-\mathbf{e}_1$. Vector $\mathbf{a}_1$ is always a radius vector of the sphere, and $\mathbf{a}_2$ and $\mathbf{a}_3$ are tangent to the sphere. The triad $\mathbf{a}_i$ of basis vectors rotates as an orthogonal unit as the surface of the sphere is traversed. For fixed r, the length of $\mathbf{a}_3$ changes with θ and vanishes at $\theta = 0°$. As r increases, both $\mathbf{a}_2$ and $\mathbf{a}_3$ increase in length. The orthogonality of the basis vectors characterizes spherical coordinates as an orthogonal space.

Further analogy with rectilinear systems permits definition of a set of contravariant (or reciprocal) basis vectors for the spherical system. These are generated by Eq. (1–26) or by Eq. (1–27), where $k = 1/r^2 \sin \theta$. Since the $\mathbf{a}_i$ are mutually orthogonal, the $\mathbf{a}^i$ are in the same directions as the $\mathbf{a}_i$ and have lengths of 1, $1/r$, and $1/(r \sin \theta)$. The reciprocal metric tensor is

$$\mathbf{g}^* = \begin{pmatrix} 1 & 0 & 0 \\ 0 & \dfrac{1}{r^2} & 0 \\ 0 & 0 & \dfrac{1}{r^2 \sin^2 \theta} \end{pmatrix} \tag{7–18}$$

It should be noted that the metric tensors are expressed in terms of the curvilinear coordinates. Cartesian coordinates were useful in defining the spherical system, but, once defined, the curvilinear system is self-sufficient.

Just as the vector product of two vectors produces an area (Section 1–8), the vector product of two spherical basis vectors leads to a differential element of area. The obvious combination is $\mathbf{a}_2 \wedge \mathbf{a}_3$, which gives an area element for fixed r (that is, on the surface of the sphere). Thus,

$$dA(\theta, \phi) = |\mathbf{a}_2 \wedge \mathbf{a}_3| \, d\theta \, d\phi = r^2 \sin\theta \, d\theta \, d\phi \tag{7–19}$$

or, in more general notation,

$$dA(\theta, \phi) = g^{1/2} a^1 \, d\theta \, d\phi \tag{7–20}$$

Integration of Eq. (7–19) over appropriate limits will give the area of portions of a spherical surface.

There are in fact two other area elements that can be defined. One of these pertains to the surface of constant θ, which is the surface of a cone of half-angle θ.

$$dA(r,\phi) = |\mathbf{a}_3 \wedge \mathbf{a}_1| \, dr \, d\phi = r \sin\theta \, dr \, d\phi \tag{7–21}$$

The other area element involves the surface of constant ϕ, which is a plane through $\mathbf{e}_3$.

$$dA(r,\theta) = |\mathbf{a}_1 \wedge \mathbf{a}_2| \, dr \, d\theta = r \, dr \, d\theta \tag{7–22}$$

EXERCISE 7–1 Integrate Eq. (7–12) to determine the volume of a sphere of radius R.

EXERCISE 7–2 A slice of thickness b is cut off a sphere of radius a, producing a dome-shaped spherical segment. Calculate the volume of the dome by integrating Eq. (7–12). [Hint: Vary r from $(a - b) \sec\theta$ to a, and vary θ from 0 to $\cos^{-1} (a - b)/a$.]

EXERCISE 7–3 Calculate the volume of a right circular cone of altitude h for which the radius of the base is R by inscribing the cone in a sphere of radius $a = (h^2 + R^2)^{1/2}$, integrating Eq. (7–12) with θ varying from 0 to $\tan^{-1} (R/h)$, and subtracting the volume of a dome of thickness $a - h$ (from Exercise 7–2).

EXERCISE 7-4 Confirm Eq. (7-17) by writing out the three basis vectors in Eq. (7-14) and taking their scalar products.

EXERCISE 7-5 Calculate $|\mathbf{g}|$ from Eq. (7-17) and show that the result is consistent with a generalization of Eq. (3-45).

EXERCISE 7-6 Find the surface area of a sphere of radius a by integrating Eq. (7-19).

EXERCISE 7-7 Find the area of the curved surface of the cone of Exercise 7-3 by integrating Eq. (7-21).

EXERCISE 7-8 Integrate Eq. (7-22) to obtain the area of the semicircle bounded by the sphere and the plane ϕ = constant.

EXERCISE 7-9 A lune is bounded on a sphere by two great circles intersecting at angle ω. Calculate the area of the lune by integrating Eq. (7-19) with ϕ varying from 0 to ω.

EXERCISE 7-10 Calculate all the derivatives $\partial \mathbf{a}_i/\partial x^j$ for spherical coordinates. Express these derivatives as linear combinations of the $\mathbf{a}_i$.

EXERCISE 7-11 By writing the transformation from cartesian basis vectors to curvilinear basis vectors in terms of coordinate derivatives, as in Eq. (3-50), prove that $\partial \mathbf{a}_i/\partial x^j = \partial \mathbf{a}_j/\partial x^i$.

EXERCISE 7-12 A spiric coordinate system with coordinates r, θ, and χ may be defined by the equations (see Hallam, 1962)

$$\begin{aligned} x &= b \cos\theta + r \cos\theta \cos\chi \\ y &= b \sin\theta + r \sin\theta \cos\chi \\ z &= r \sin\chi \end{aligned}$$

(a) Calculate matrices **F** and **G** for the transformation.
(b) Calculate the Jacobian of the transformation.
(c) Determine the spiric basis vectors $\mathbf{a}_i$ in terms of the cartesian $\mathbf{e}_i$.
(d) Calculate the metric tensors $\mathbf{g}$ and $\mathbf{g}^*$.
(e) Describe the solid generated by varying r from 0 to a (where $a < b$), θ from 0 to 2π, and χ from 0 to 2π, and calculate the volume of this solid.
(f) Calculate the area of the surface defined by $r = a$.

EXERCISE 7-13 Parabolic coordinates ξ, η, and ϕ are defined by $x = \sqrt{\xi\eta}\cos\phi$, $y = \sqrt{\xi\eta}\sin\phi$, $z = \frac{1}{2}(\xi - \eta)$. Calculate $\mathbf{g}$, $\mathbf{g}^*$, J, and dV for parabolic coordinates.

EXERCISE 7-14 A coordinate system with variables r, θ, and s is defined in terms of cartesian coordinates x, y, z by $x = r\cos\theta$, $y = r\sin\theta$, $z = s\theta$.

(a) Determine the metric tensors $\mathbf{g}$ and $\mathbf{g}^*$ and show that the system is not orthogonal.

(b) Calculate the differential volume element.

7-3 Derivatives of Tensors

One of the most useful operators involving derivatives is the del operator ∇, which is defined in a cartesian coordinate system with basis vectors $\mathbf{e}_i$ and components l^i as

$$\nabla = \mathbf{e}_i \frac{\partial}{\partial l^i} \tag{7-23}$$

$$\nabla_i = \frac{\partial}{\partial l^i} \tag{7-24}$$

When ∇ operates on a scalar function ψ, the gradient is produced; in cartesian coordinates this is

$$\textbf{grad}\ \psi = \nabla\psi = \mathbf{e}_i \frac{\partial \psi}{\partial l^i} \tag{7-25}$$

When ∇ operates on a vector, the divergence is produced; the cartesian formula for the divergence is

$$\text{div}\ \mathbf{v} = \nabla \cdot \mathbf{v} = \frac{\partial v^i}{\partial l^i} \tag{7-26}$$

The curl of a vector $\mathbf{v}$ is defined as $\nabla \wedge \mathbf{v}$, and the cartesian form of this is

$$\text{curl}\ \mathbf{v} = \nabla \wedge \mathbf{v} = \begin{vmatrix} \mathbf{e}_1 & \mathbf{e}_2 & \mathbf{e}_3 \\ \dfrac{\partial}{\partial l^1} & \dfrac{\partial}{\partial l^2} & \dfrac{\partial}{\partial l^3} \\ v^1 & v^2 & v^3 \end{vmatrix} \tag{7-27}$$

Finally, the very important Laplacian operator ∇^2 has the cartesian form

$$\nabla^2 = \frac{\partial^2}{\partial l^i \partial l^i} \tag{7-28}$$

The desirable extension of these results to general coordinate systems is complicated by the fact that derivatives of tensors are not necessarily tensors. The derivative of a scalar with respect to a contravariant coordinate is a covariant vector, for by the rules of differentiation

$$\frac{\partial \psi}{\partial x'^i} = \frac{\partial \psi}{\partial x^j} \frac{\partial x^j}{\partial x'^i} \tag{7-29}$$

which has the form of the covariant transformation law [see Eq. (3-62)]. Likewise, $\partial\psi/\partial x_i$ produces a contravariant vector. However, such derivatives as $\partial v^k/\partial x^l$ do not transform as tensors. In this case

$$\frac{\partial v'^i}{\partial x'^j} = \left(\frac{\partial}{\partial x'^j} \frac{\partial x'^i}{\partial x^k} v^k \right) \tag{7-30}$$

which may be expanded to

$$\frac{\partial v'^i}{\partial x'^j} = \frac{\partial^2 x'^i}{\partial x'^j \, \partial x^k} v^k + \frac{\partial x'^i}{\partial x^k} \frac{\partial x^l}{\partial x'^j} \frac{\partial v^k}{\partial x^l} \tag{7-31}$$

Without the first term on the right-hand side of Eq. (7-31), $\partial v^k/\partial x^l$ would transform as a second-rank mixed tensor with one contravariant and one covariant suffix. Within purely rectilinear systems this term does vanish, and derivatives of vectors and tensors are tensors, but this is not true in curvilinear systems. The case of $\partial w_i/\partial x^k$ was analyzed in a similar fashion in Section 3-10.

The complication in these derivatives arises because the basis vectors in curvilinear systems are not constant. Thus, the derivative of a vector must include the derivatives of the $\mathbf{a}_i$:

$$d\mathbf{v} = \mathbf{a}_i \, dv^i + v^i \frac{\partial \mathbf{a}_i}{\partial x^j} dx^j \tag{7-32}$$

The partial derivatives of the basis vectors are related to the basis vectors by coefficients $\left\{ {k \atop i \;\; j} \right\}$ defined by

$$\frac{\partial \mathbf{a}_i}{\partial x^j} = \begin{Bmatrix} k \\ i\ j \end{Bmatrix} \mathbf{a}_k \tag{7-33}$$

Thus,

$$d\mathbf{v} = \mathbf{a}_i\, dv^i + \begin{Bmatrix} k \\ i\ j \end{Bmatrix} v^i \mathbf{a}_k\, dx^j \tag{7-34}$$

By changing the dummy indices,

$$d\mathbf{v} = (dv^i + \begin{Bmatrix} i \\ k\ l \end{Bmatrix} v^k\, dx^l)\mathbf{a}_i \tag{7-35}$$

Dividing by dx^l and keeping the other dx^j constant leads to

$$\frac{\partial \mathbf{v}}{\partial x^l} = v^i_{;l}\ \mathbf{a}_i \tag{7-36}$$

where

$$v^i_{;l} = \frac{\partial v^i}{\partial x^l} + \begin{Bmatrix} i \\ k\ l \end{Bmatrix} v^k \tag{7-37}$$

The quantity $v^i_{;l}$ is known as the *covariant derivative* of the vector with components v^i. Similarly,

$$v_{i;l} = \frac{\partial v_i}{\partial x^l} - \begin{Bmatrix} k \\ i\ l \end{Bmatrix} v_k \tag{7-38}$$

is the covariant derivative of the vector with components v_i.

The scalar product of Eq. (7–33) with $g^{ml}\mathbf{a}_l$ yields

$$\begin{Bmatrix} m \\ \mathrm{i}\ \ \mathrm{j} \end{Bmatrix} = g^{ml}\mathbf{a}_l \cdot \frac{\partial \mathbf{a}_i}{\partial x^j} = \mathbf{a}^m \cdot \frac{\partial \mathbf{a}_i}{\partial x^j} \tag{7-39}$$

It is convenient to introduce the symbol $[ij, l]$:

$$[ij,l] = \mathbf{a}_l \cdot \frac{\partial \mathbf{a}_i}{\partial x^j} \tag{7-40}$$

The quantities $[ij, l]$ and $\left\{ {k \atop i\ j} \right\}$ are known as *Christoffel symbols*, of the first kind and second kind, respectively. Alternative expressions for the Christoffel symbols are

$$[ij, l] = \frac{1}{2}\left(\frac{\partial g_{il}}{\partial x^j} + \frac{\partial g_{jl}}{\partial x^i} - \frac{\partial g_{ij}}{\partial x^l}\right) \tag{7-41}$$

$$\left\{ {m \atop i\ j} \right\} = \frac{1}{2} g^{ml}\left(\frac{\partial g_{il}}{\partial x^j} + \frac{\partial g_{jl}}{\partial x^i} - \frac{\partial g_{ij}}{\partial x^l}\right) \tag{7-42}$$

The Christoffel symbols vanish in all rectilinear coordinate systems, for which the metric tensor elements are constants.

The Christoffel symbols are useful in creating tensor generalizations of the various differential operators. Because $\partial\psi/\partial x^i$ is a covariant tensor of rank one, the gradient given by Eq. (7-25) may be transformed to curvilinear coordinates to produce

$$\mathbf{grad}\ \psi = \nabla\psi = \frac{\partial x^j}{\partial l^i}\frac{\partial x^k}{\partial l^i}\mathbf{a}_j\frac{\partial\psi}{\partial x^k} \tag{7-43}$$

By Eq. (3-60),

$$\nabla\psi = g^{jk}\mathbf{a}_j\frac{\partial\psi}{\partial x^k} = \mathbf{a}^k\frac{\partial\psi}{\partial x^k} \tag{7-44}$$

In spherical coordinates the gradient is

$$\nabla\psi = \mathbf{a}_1\frac{\partial\psi}{\partial r} + \frac{\mathbf{a}_2}{r^2}\frac{\partial\psi}{\partial\theta} + \frac{\mathbf{a}_3}{r^2\sin^2\theta}\frac{\partial\psi}{\partial\phi} \tag{7-45}$$

Generalization of the divergence of a vector is obtained by contraction of the covariant derivative given in Eq. (7-35).

$$\operatorname{div} \mathbf{v} = \nabla \cdot \mathbf{v} = v^i_{;i} = \frac{\partial v^i}{\partial x^i} + \left\{ \begin{matrix} i \\ k \ i \end{matrix} \right\} v^k \tag{7-46}$$

Expressing the Christoffel symbols in terms of the metric tensor components leads eventually (see, for example, Spiegel, 1959, or Hay, 1953) to the compact and convenient formula

$$\operatorname{div} \mathbf{v} = \frac{1}{g^{1/2}} \frac{\partial}{\partial x^i} (g^{1/2} v^i) \tag{7-47}$$

In spherical coordinates the divergence is

$$\operatorname{div} \mathbf{v} = \frac{\partial v^1}{\partial r} + \frac{\partial v^2}{\partial \theta} + \frac{\partial v^3}{\partial \phi} + \frac{2}{r} v^1 + v^2 \cot \theta \tag{7-48}$$

The curl of a vector $\mathbf{v}$ in a generalized space is

$$\nabla \Lambda \mathbf{v} = -\epsilon^{ijk} v_{j,k} \mathbf{a}_i \tag{7-49}$$

where $v_{j,k}$ is the covariant derivative given by Eq. (7-38), and ϵ^{ijk} is the permutation tensor defined in Section 1-14. Inserting the values of ϵ^{ijk},

$$\nabla \Lambda \mathbf{v} = \frac{1}{g^{1/2}} [\mathbf{a}_1(v_{3,2} - v_{2,3}) + \mathbf{a}_2(v_{1,3} - v_{3,1}) + \mathbf{a}_3(v_{2,1} - v_{1,2})] \tag{7-50}$$

For a spherical coordinate system

$$\nabla \Lambda \mathbf{v} = \frac{1}{r^2 \sin \theta} \left\{ \left[2r^2 \sin \theta \cos \theta \, v^3 + (r \sin \theta)^2 \frac{\partial v^3}{\partial \theta} - r^2 \frac{\partial v^2}{\partial \phi} \right] \mathbf{a}_1 + \left[\frac{\partial v^1}{\partial \phi} - 2r \sin^2\theta v^3 - (r \sin \theta)^2 \frac{\partial v^3}{\partial r} \right] \mathbf{a}_2 + \left[2rv^2 + r^2 \frac{\partial v^2}{\partial r} - \frac{\partial v^1}{\partial \theta} \right] \mathbf{a}_3 \right\} \tag{7-51}$$

The Laplacian of a scalar function ψ may be determined directly by taking the divergence of the gradient of ψ. Combining Eqs. (7-44) and (7-47) produces

$$\nabla^2\psi = \frac{1}{g^{1/2}}\frac{\partial}{\partial x^i}\left(g^{1/2}g^{ij}\frac{\partial\psi}{\partial x^j}\right) \tag{7-52}$$

In spherical coordinates the Laplacian is

$$\nabla^2\psi = \frac{1}{r^2}\frac{\partial}{\partial r}\left(r^2\frac{\partial\psi}{\partial r}\right)+\frac{1}{r^2\sin\theta}\frac{\partial}{\partial\theta}\left(\sin\theta\frac{\partial\psi}{\partial\theta}\right)+\frac{1}{r^2\sin^2\theta}\frac{\partial^2\psi}{\partial\phi^2} \tag{7-53}$$

EXERCISE 7-15 (a) Calculate all the Christoffel symbols (of both kinds) for spherical coordinates.

(b) Calculate all the Christoffel symbols (of both kinds) for spiric coordinates (see Exercise 7-12).

EXERCISE 7-16 Calculate div **v** for a rectilinear system.

EXERCISE 7-17 Evaluate all the quantities in Eq. (7-50) for a spherical coordinate system, and hence derive Eq. (7-51).

EXERCISE 7-18 Evaluate all the Christoffel symbols for the parabolic coordinate system of Exercise 7-13.

EXERCISE 7-19 Evaluate all the Christoffel symbols for the coordinate system of Exercise 7-14.

EXERCISE 7-20 Obtain the divergence in spherical coordinates from Eq. (7-46) by using the Christoffel symbols from Exercise 7-15.

EXERCISE 7-21 Derive Eq. 7-53 for the Laplacian in spherical coordinates.

EXERCISE 7-22 Derive expressions for the gradient, divergence, curl, and Laplacian in the spiric coordinate system of Exercise 7-12.

EXERCISE 7-23 Derive expressions for the gradient, divergence, curl, and Laplacian in the parabolic coordinate system of Exercise 7-13.

EXERCISE 7-24 Derive expressions for the gradient, divergence, curl, and Laplacian for the coordinate system of Exercise 7-14.

EXERCISE 7–25 Derive expressions for the gradient, divergence, curl, and Laplacian for a general rectilinear coordinate system (a triclinic crystal).

EXERCISE 7–26 Generalized momentum may be defined as $p_i = \partial T/\partial \dot{q}^i$ where T is the kinetic energy, and $\dot{\mathbf{q}}$ is the time derivative of the positional coordinate $\mathbf{q}$. If $\dot{\mathbf{q}}$ transforms the same way as $\mathbf{q}$, what is the tensor character of the momentum? What may be said then about products such as $p_i q^i$ that appear in the Heisenberg uncertainty principle, or about the differential products $dp_i\, dq^i$ that give the volume elements in Gibbs phase space?

References

Aris, R. (1962). *Vectors, Tensors, and the Basic Equations of Fluid Mechanics.* Prentice-Hall, Englewood Cliffs, New Jersey.

Ayres, F. (1962). *Theory and Problems of Matrices.* Schaum Publishing Co., New York, New York.

Barrett, C. S. (1952). *Structure of Metals,* 2nd edition. McGraw-Hill, New York, New York.

Barsch, G. R. (1976). X-Ray Determination of Piezoelectric Constants. *Acta Cryst.,* A32, 575–586.

Beguemsi, T., Garnier, P., and Weigel, D. (1978). Evolution des Tenseurs de Dilatation Thermique en Fonction de la Temperature. II. *J. Solid State Chem.,* 25, 315–324.

Billings, A. R. (1969). *Tensor Properties of Materials.* Wiley–Interscience, New York, New York.

Buerger, M. J. (1942). *X-Ray Crystallography.* John Wiley and Sons, New York, New York.

Buerger, M. J. (1944). *The Photography of the Reciprocal Lattice.* American Society for X-Ray and Electron Diffraction, Monograph No. 1.

Buerger, M. J. (1964). *The Precession Method.* John Wiley and Sons, New York, New York.

Busing, W. R., and Levy, H. A. (1958). Determination of the Principal Axes of the Anisotropic Temperature Factor. *Acta Cryst.,* 11, 450–451.

Carslaw, H. S., and Jaeger, J. C. (1947). *Conduction of Heat in Solids.* Oxford University Press, London.

Cornwell, J. F. (1969). *Group Theory and Electronic Energy Bands in Solids.* North-Holland Publishing Co., Amsterdam–London. Distributed by American Elsevier Publishing Co.

Cruickshank, D. W. J. (1956). The Analysis of the Anisotropic Thermal Motion of Molecules in Crystals. *Acta Cryst.,* 9, 754–756.

Enraf-Nonius Company (1977). *Manual for CAD-4 Kappa Diffractometer.* Delft, Holland.

Flügge, W. (1972). *Tensor Analysis and Continuum Mechanics.* Springer-Verlag, New York, Heidelberg, Berlin.

Fumi, F. G., and Ripamonti, C. (1980). Tensor Properties and Rotational Symmetry of Crystals. *Acta Cryst.,* A36, 535–551, 551–558.

Gaylord, T. K. (1975). Tensor Description of Physical Properties of Crystals. *Amer. J. Phys.,* 43, 861–868.

Goldstein, H. (1980). *Classical Mechanics,* 2nd edition. Addison-Wesley Publishing Co., Reading, Mass.

Hallam, A. F. (1962). On a Curvilinear Coordinate System. *Amer. Math. Monthly,* 69, 105–113.

Hamilton, W. C. (1959). On the Isotropic Temperature Factor Equivalent to a Given Anisotropic Temperature Factor. *Acta Cryst.,* 12, 609–610.

Hamilton, W. C. (1974). Angle Settings for Four-Circle Diffractometers. *International Tables for X-Ray Crystallography,* Vol. IV, Chapter 3. The Kynoch Press, Birmingham, England.

Haussühl, S. (1977). Elastic, Thermoelastic, and Dynamic Piezoelectric Properties of Trigonal Potassium Bromate. *Acta Cryst.,* A33, 320–322.

Hay, G. E. (1953). *Vector and Tensor Analysis.* Dover Publications, New York, New York.

Ho, C. Y., Powell, R. W., and Liley, P. E. (1972). Thermal Conductivity of the Elements. *J. Phys. Chem. Ref. Data,* 1, 279–421.

Huntington, H. B. (1958). *The Elastic Constants of Crystals.* Reprinted from *Solid State Physics,* Vol. 7. Academic Press, New York, New York.

International Tables for X-Ray Crystallography. Volume I, (1952) Symmetry Groups. Volume II, (1959) Mathematical Tables. Volume III, (1962) Physical and Chemical Tables. Volume IV, (1974) Revised and Supplementary Tables. The Kynoch Press, Birmingham, England.

Johnson, C. K. (1970). An Introduction to Thermal Motion Analysis. Topic Fl in *Crystallographic Computing,* edited by F. R. Ahmed. Munksgaard, Copenhagen.

Johnson, C. K., and Levy, H. A. (1974). Thermal-Motion Analysis Using Bragg Diffraction Data. *International Tables for X-Ray Crystallography,* Vol. IV, Chapter 5. The Kynoch Press, Birmingham, England.

Kittel, C. (1971). *Introduction to Solid State Physics,* 4th edition. John Wiley and Sons, New York, New York.

Koster, G. F. (1957). *Space Groups and Their Representations.* Academic Press, New York and London.

Kumaraswamy, K., and Krishnamurthy, N. (1980). The Acoustic Gyrotropic Tensor in Crystals. *Acta Cryst.,* A36, 760–762.

McConnell, A. J. (1957). *Applications of Tensor Analysis.* Reprinting of 1931 edition. Dover Publications, New York, New York.

Miller, D. G. (1960). Thermodynamics of Irreversible Processes, The Experimental Verification of the Onsager Reciprocal Relations. *Chem Rev.,* 60, 15–37, 593.

Neustadt, R. J., Cagle, F. W., and Waser, J. (1968). Vector Algebra and the Relations between Direct and Reciprocal Lattice Quantities. *Acta Cryst.,* A24, 247–248.

Nye, J. F. (1957). *Physical Properties of Crystals.* Oxford University Press, London.

Patterson, A. L. (1959). Vector and Tensor Analysis. *International Tables for X-Ray Crystallography,* Vol. II, Chapter 2. The Kynoch Press, Birmingham, England.

Pawley, G. S. (1968). Anisotropic Temperature Factors and Screw Rotation Coefficients from a Lattice Dynamical Viewpoint. *Acta Cryst.,* B24, 485–486.

Portigal, D. L., and Burstein, E. (1968). Acoustical Activity and Other First-Order Spatial Dispersion Effects in Crystals. *Phys. Rev.,* 170, 673–678.

Sadanandam, J., and Suryanarayana, S. V. (1979). Thermal Expansion of α-NH_4HgCl_3. *Acta Cryst.,* A35, 923-924.

Sands, D. E. (1966). Transformations of Variance–Covariance Tensors. *Acta Cryst.,* 21, 868–872.

Sands, D. E. (1969). *Introduction to Crystallography.* W. A. Benjamin, Addison-Wesley Publishing Co., Reading, Mass.

Schomaker, V., and Trueblood, K. N. (1968). On the Rigid-Body Motion of Molecules in Crystals. *Acta Cryst.,* B24, 63–76.

Seitz, F. A. (1934–1936). A Matrix-Algebraic Development of the Crystallographic Groups. *Z. Kristallogr.* 88, 433–459; 90, 289–313; 91, 336–366; 94, 100–130.

Shmueli, U. (1974). On the Standard Deviation of Dihedral Angle. *Acta Cryst.,* A30, 848–849.

Simmons, G., and Wang, H. (1971). *Single Crystal Elastic Constants and Calculated Aggregate Properties: A Handbook,* 2nd edition. MIT Press, Cambridge, Mass.

Smith, C. S. (1958). Macroscopic Symmetry and Properties of Crystals. *Solid State Physics,* Vol. 7. Academic Press, New York, New York.

Spiegel, M. R. (1959). *Vector Analysis and an Introduction to Tensor Analysis.* Schaum Publishing Co., New York, New York.

Waser, J. (1951). The Lorentz Factor for the Buerger Precession Method. *Rev. Sci. Instrum.,* 22, 563–568.

Waser, J. (1955). The Anisotropic Temperature Factor in Triclinic Coordinates. *Acta Cryst.,* 8, 731.

Weigel, D., Beguesmi, T., Garnier, P., and Berar, J. F. (1978). Evolution des Tenseurs de Dilatation Thermique en Fonction de la Temperature. I. *J. Solid State Chem.,* 23, 241–251.

Zachariasen, W. H. (1945). *Theory of X-Ray Diffraction in Crystals.* John Wiley and Sons, New York, New York.

Solutions to Exercises

1-2 [1.25, 0.75], [−0.25, 1.25].
1-3 22.5 Å^2.
1-5 9.80 Å, 2.90 Å.
1-6 11.10 Å, 9.25 Å.
1-7 57.6°.
1-8 31.5°.
1-10 $\cos^{-1}(-1/3) = 109.47°$ = tetrahedral angle.
1-11 90°.
1-12 120.00°.
1-13 $[1/\sqrt{2}, -1/\sqrt{2}, 0]$.
1-14 (a) 6.76 Å; (b) 6.44 Å; (c) 126.4°; (d) [−0.300, −1.000, 0.700], 11.78 Å.
1-15 82.1°, 132.2°, 14.9°.
1-17 $g_{11} = 25.0$, $g_{12} = 9.06$, $g_{13} = 3.92$, $g_{22} = 49.0$, $g_{23} = -31.5$, $g_{33} = 81.0$.
1-18 $a_1 = 7.56$, $a_2 = 8.66$, $a_3 = 10.49$, $\alpha = 90.0°$, $\beta = 93.0°$, $\gamma = 90.0°$.
1-20 88.0°.
1-21 100.2°.
1-22 1.78 Å.
1-23 Each Sc atom has six F neighbors at the vertices of a slightly irregular octahedron; Sc-F = 2.017 Å. Each F atom is shared by two octahedra and is bonded to two Sc atoms. Each F atom has four F neighbors in each octahedron—one at 2.712 Å, one at 3.015 Å, and two at 2.846 Å. Of the fifteen F-Sc-F angles within an octahedron, three are 84.5°, six are 89.7°, three are 96.7°, and three are 171.3°.

1-24 $\begin{pmatrix} 0 & \mathbf{e}_3 & -\mathbf{e}_2 \\ -\mathbf{e}_3 & 0 & \mathbf{e}_1 \\ \mathbf{e}_2 & -\mathbf{e}_1 & 0 \end{pmatrix}$.

1-26 (a) $\mathbf{u} \cdot (\mathbf{u} \wedge \mathbf{v}) = 0$; (b) $\mathbf{u} \cdot (\mathbf{v} \wedge \mathbf{w}) = 0$ if and only if the vectors are coplanar (or if any one of them is the null vector of length 0).

1-27 283 Å^3.

1-28 (a) $a_1a_2a_3 \sin \beta$; (b) $a_1a_2a_3$; (c) $(a_1)^2a_3$; (d) $(a_1)^2a_3 \sqrt{3}/2$; (e) $(a_1)^3 (1 - \cos \alpha) \times (1 + 2 \cos \alpha)^{1/2}$; (f) $(a_1)^3$.

1-29 431 Å^3.

1-30 $a^1 = 0.210$, $a^2 = 0.1429$, $a^3 = 0.1237$ Å^{-1}, $\beta^* = 72.0°$.

1-32 $a^i = 1.000$, $\alpha^* = \beta^* = \gamma^* = 90°$.

1-33 (a) $\mathbf{g} = \begin{pmatrix} 64.0 & -6.97 & -24.8 \\ -6.97 & 100.0 & -20.8 \\ -24.8 & -20.8 & 144.0 \end{pmatrix}$, (b) $\mathbf{g}^* = \begin{pmatrix} 0.01708 & 0.00186 & 0.00322 \\ 0.00186 & 0.01051 & 0.00184 \\ 0.00322 & 0.00184 & 0.00777 \end{pmatrix}$.

(c) $a^1 = 0.1307$, $a^2 = 0.1025$, $a^3 = 0.0881$ Å^{-1}.
$\alpha^* = 78.2°$, $\beta^* = 73.8°$, $\gamma^* = 82.0°$.

1-34 $a^1 = 0.239$, $a^2 = 0.1761$, $a^3 = 0.1191$ Å^{-1}, $\alpha^* = 110.5°$, $\beta^* = 105.0°$, $\gamma^* = 57.1°$.

1-35 3.

1-36 (a) 3; (b) 3; (c) 3.

1-37 From $\bar{\mathbf{u}}\mathbf{g}\mathbf{u} = 1$, and $\mathbf{u} = \mathbf{g}^*\mathbf{v}$ where $v^I = a_I \cos \chi_I$ (no summation), it follows that $\bar{\mathbf{v}}\mathbf{g}^*\mathbf{v} = 1$, which is the desired relationship.

1-38 $g = 6.48 \times 10^4$, $g^* = 1.544 \times 10^{-5}$, $V = 254.5$ Å^3, $V^* = 3.93 \times 10^{-3}$ Å^{-3}.

1-39 $g = 8.01 \times 10^4$, $g^* = 1.249 \times 10^{-5}$, $V = 283$ Å^3, $V^* = 3.53 \times 10^{-3}$ Å^{-3}.

1-40 $g = 1.859 \times 10^5$, $g^* = 5.38 \times 10^{-6}$, $V = 431$ Å^3, $V^* = 2.32 \times 10^{-3}$ Å^{-3}.

1-41 $g = 8.18 \times 10^5$, $g^* = 1.223 \times 10^{-6}$, $V = 904$ Å^3, $V^* = 1.106 \times 10^{-3}$ Å^{-3}.

1-44 $a_1 = 5.00$, $a_2 = 6.50$, $a_3 = 7.10$ Å, $\alpha = 98.0°$, $\beta = 101.5°$, $\gamma = 95.6°$, $a^1 = 0.206$, $a^2 = 0.1567$, $a^3 = 0.1456$ Å^{-1}, $\alpha^* = 80.7°$, $\beta^* = 77.5°$, $\gamma^* = 82.6°$.

1-45 14.6 Å.

1-49 6.

1-51 $g^{ij}\epsilon_{ipq}\epsilon_{jrs} = g_{pr}g_{qs} - g_{ps}g_{qr}$.

1-57 8.21×10^3 Å^3.

1-59 $(\mathbf{u} \wedge \mathbf{v}) \wedge \mathbf{w} = -(\mathbf{v} \cdot \mathbf{w})\mathbf{u} + (\mathbf{u} \cdot \mathbf{w})\mathbf{v}$.

1-60 **0**.

1-62 [0.0878, 0.0444, 0.124].

1-63 (a) S – O(1) = 1.450, S – O(2) = 1.443, C(1) – O(3) = 1.442, C(2) – O(3) = 1.437 Å; (b) O(1) – S – O(2) = 117.4°, C(1) – O(3) – C(2) = 60.4°; (c) [–0.303, 0.132, 0.270]; (d) 90.1°.

2-2
$$\begin{aligned} 0.320x^2 + 0.150x^3 &= 0.0320 \\ 0.320x^1 + 0.070x^3 &= 0.0704 \\ 0.150x^1 - 0.070x^2 &= 0.0260. \end{aligned}$$

2-3
$$\begin{pmatrix} 0 & v^3 & -v^2 \\ -v^3 & 0 & v^1 \\ v^2 & -v^1 & 0 \end{pmatrix} \begin{pmatrix} x^1 \\ x^2 \\ x^3 \end{pmatrix} = \begin{pmatrix} u^2v^3 - u^3v^2 \\ u^3v^1 - u^1v^3 \\ u^1v^2 - u^2v^1 \end{pmatrix}.$$

2-4 $x^2 = -0.200$, $0.160x^1 + 0.100x^3 = 0.0265$.

2-5 $-1.392x^1 + 6.762x^2 + 8.451x^3 = -0.557$; $6.815x^1 - 0.729x^2 - 8.315x^3 = 2.726$.

2-6 $[-0.023, -0.078, -0.058]$.

2-9 $15x^1 + 9x^2 - 53x^3 = -3.8$.

2-10 $-13.22x^1 + 18.00x^2 + 32.00x^3 = -3.78$.

2-11 (a) (230); (b) (½, 1, 5), which is parallel to the lattice plane (1, 2, 10); (c) (362); (d) (3/2, 3, 1), which is parallel to the lattice plane (362).

2-12 $\mathbf{z} = V(v^2w^3, u^1w^3, u^1v^2)$; $\mathbf{z} \wedge \mathbf{h} = \mathbf{0}$.

2-13 Indices are $(-3.9, -2.4, 13.9)$. Intercepts are $[-0.253, 0, 0]$, $[0, -0.422, 0]$, $[0, 0, 0.072]$.

2-14 $-6x^1 + 15x^2 + 14x^3 = 26$. Intercepts are $[-26/6, 0, 0]$, $[0, 26/15, 0]$, $[0, 0, 26/14]$. Indices are $(-6/26, 15/26, 14/26)$, which is a plane parallel to the lattice plane $(-6, 15, 14)$.

2-15 $h_i = g_{ij}v^j/(\mathbf{u} \cdot \mathbf{v})$.

2-16 Intercepts are $[0.286, 0, 0]$, $[0, -0.210, 0]$, $[0, 0, -0.118]$. Indices are $(3.50, -4.76, -8.46)$.

2-17 $h = 1.322$ Å^{-1}; distance to plane = 0.756 Å.

2-18 0.559 Å.

2-19 $\mathbf{h} = (-3.95, -2.37, 13.95)$, $\mathbf{u} \cdot \mathbf{h} = 2.76$; distance to plane = 3.14 Å.

2-21 $2x^1 + 4x^2 + 6x^3 = 5$. Since the left-hand side of this equation is an even integer for integer values of x^1, x^2, x^3 and the right-hand side is odd, this plane does not pass through lattice points of a primitive unit cell.

2-22 $d(210) = 2.64$ Å; $d(111) = 3.20$ Å.

2-23 (a) Monoclinic:
$$\left(\frac{1}{d}\right)^2 = \left(\frac{h_1}{a_1 \sin\beta}\right)^2 + \left(\frac{h_2}{a_2}\right)^2 + \left(\frac{h_3}{a_3 \sin\beta}\right)^2 - \frac{2h_1h_3\cos\beta}{a_1a_3\sin^2\beta}.$$

(b) Orthorhombic:
$$\left(\frac{1}{d}\right)^2 = \left(\frac{h_1}{a_1}\right)^2 + \left(\frac{h_2}{a_2}\right)^2 + \left(\frac{h_3}{a_3}\right)^2.$$

(c) Tetragonal:
$$\left(\frac{1}{d}\right)^2 = \frac{(h_1)^2 + (h_2)^2}{(a_1)^2} + \frac{(h_3)^2}{(a_3)^2}.$$

(d) Hexagonal:
$$\left(\frac{1}{d}\right)^2 = \frac{4[(h_1)^2 + h_1h_2 + (h_2)^2]}{3(a_1)^2} + \left(\frac{h_3}{a_3}\right)^2.$$

(e) Rhombohedral:

$$\left(\frac{1}{d}\right)^2 = \frac{[(h_1)^2 + (h_2)^2 + (h_3)^2](1 + \cos\alpha) - 2[h_1h_2 + h_2h_3 + h_3h_1]\cos\alpha}{(a_1)^2\,(1 + \cos\alpha - 2\cos^2\alpha)}.$$

(f) Cubic: $$\left(\frac{1}{d}\right)^2 = \frac{(h_1)^2 + (h_2)^2 + (h_3)^2}{(a_1)^2}.$$

2–24 (a) 41.4°; (b) 90.0°; (c) 53.9°; (d) 74.7°; (e) 36.4°.
2–25 154.6°.
2–26 (a) 109.5°; (b) 144.7°; (c) 90.0°; (d) 125.3°; (e) 140.8°.
2–28 $\mathbf{P}(\mathbf{v}\|\mathbf{h})$ = [0.061, −0.008, −0.123] = (5.41, −0.73, −9.02); length = 1.201 Å. $\mathbf{P}(\mathbf{w}\|\mathbf{h})$ = [−0.044, 0.182, 0.089] = (−5.21, 6.86, 6.02); length = 1.419 Å. Angle between projected vectors = 130.4°; angle between **v** and **w** = 120.2°.
2–29 (−1, 3, 0.274).
2–30 **h** = [0, 0, 1] = (−16.14, 0, 72.4), h = 8.51 Å, **u** = ($\bar{1}$, 3, 2), **h** • **u** = 2, $\mathbf{P}(\mathbf{u}\|\mathbf{h})$ = (−0.554, 3, 0).
2–31 (a) (2, 1, −0.547); (b) **h** •**u** = 0, so the projected vector is (210).

3–1 (a) 1.42 Å; (b) [1.150, 0.120, −2.065], 20.7 Å.
3–2 1.97 Å.
3–3 [±1/2, ±1/2, 0], [±1/2, 0, ±1/2], [0, ±1/2, ±1/2].

3–4 $$\mathbf{F} = \begin{pmatrix} ^2/_5 & ^3/_5 & ^1/_3 \\ ^1/_5 & -^1/_5 & 0 \\ 0 & 0 & ^1/_3 \end{pmatrix}.$$

x^1	x^2	x^3	x'^1	x'^2	x'^3
1	0	0	$^2/_5$	$^1/_5$	0
0	1	0	$^3/_5$	$-^1/_5$	0
0	0	1	$^1/_3$	0	$^1/_3$
1	1	1	$^4/_3$	0	$^1/_3$
1	1	0	1	0	0
3	−2	0	0	1	0
−1	−1	3	0	0	1
4	−1	0	1	1	0

3–5 (a) $a_1' = a_2' = a_3'$ = 6.46 Å, $\alpha' = \beta' = \gamma'$ = 83.0°; dimensions suggest rhombohedral system.

(b) a_1' = 8.56, a_2' = 8.57, a_3' = 12.48 Å, α' = 90.0°, β' = 90.0°, γ' = 120.0°; dimensions suggest hexagonal system. Volume is 3/2 that of *C*-centered cell, so there are three lattice points per unit cell.

(c) $[x^1, x^2, x^3] = \pm[0.167, 0, 0.333]$; $[x'^1, x'^2, x'^3] = \pm[0.167, 0.167, -0.333]$.

(d)

x^1	x^2	x^3	x'^1	x'^2	x'^3
±[0.167,	0.000,	0.333]	±[0.167,	0.333,	0.000]
±[0.667,	0.500,	0.333]	±[0.833,	0.667,	0.333]
±[0.667,	0.500,	−0.667]	±[0.500,	0.000,	0.667]

3–6 (a) $a_1' = a_2' = 5.670$, $a_3' = 7.020$ Å. Sc atoms at 0, 0, 0; 2/3, 1/3, 1/3; 1/3, 2/3, 2/3. F atoms at 0.333, 0.194, 0.167; 0, 0.527, 0.500; 0.666, 0.861, 0.834; −0.194, 0.140, 0.167; 0.473, 0.473, 0.500; 0.139, 0.807, 0.834; −0.140, −0.333, 0.167; 0.527, 0.000, 0.500; 0.193, 0.334, 0.834.

3–7 (a) $a_1' = 6.24$, $a_2' = 25.1$, $a_3' = 25.4$ Å, $\alpha' = 92.0°$, $\beta' = 82.2°$, $\gamma' = 105.0°$.
(b) $g_{11} = 39.0$, $g_{22} = 631$, $g_{33} = 647$, $g_{12} = -40.5$, $g_{13} = 21.7$, $g_{23} = -22.0$.
(d) $g^{11} = 0.0280$, $g^{22} = 0.00170$, $g^{33} = 0.00158$, $g^{12} = 0.00176$, $g^{13} = -0.00088$, $g^{23} = 0.0000$.

3–8 $g_{11}' = 72.0$, $g_{22}' = 468$, $g_{33}' = 648$, $g_{12}' = 36.0$, $g_{13}' = -72.0$, $g_{23}' = -36.0$.

3–10

x^1	x^2	x^3	x'^1	x'^2	x'^3	x''^1	x''^2	x''^3
0.50,	0.50,	0.00	0.25,	−0.50,	0.50	0.00,	0.00,	0.25
0.00,	0.00,	0.00	0.00,	0.00,	0.00	0.00,	0.00,	0.00
3.00,	1.00,	1.00	1.00,	−1.00,	0.00	4.00,	−1.00,	−3.00
11.50,	7.50,	−0.50	3.50,	−2.50,	3.00	3.00,	0.50,	0.50

3–11

$$\mathbf{g}' = \begin{pmatrix} 20 & 6 & 2 \\ 6 & 3 & 2 \\ 2 & 2 & 2 \end{pmatrix}, \qquad \mathbf{g}'' = \begin{pmatrix} 13 & 11 & 10 \\ 11 & 11 & 8 \\ 10 & 8 & 8 \end{pmatrix}.$$

$V' = 2, \qquad V'' = 2.$

3–12 $a''^1 = 2.45$, $a''^2 = 1.00$, $a''^3 = 2.35$ Å, $\alpha''^* = 50.2°$, $\beta''^* = 163.2°$, $\gamma''^* = 144.7°$.

3–13

$$\mathbf{G} = \begin{pmatrix} 1/a_1 & 0 & 0 \\ -\cot\gamma/a_1 & 1/(a_2\sin\gamma) & 0 \\ a^1\cos\beta^* & a^2\cos\alpha^* & a^3 \end{pmatrix}, \quad \mathbf{F} = \begin{pmatrix} a_1 & a_2\cos\gamma & a_3\cos\beta \\ 0 & a_2\sin\gamma & -a_3\sin\beta\cos\alpha^* \\ 0 & 0 & 1/a^3 \end{pmatrix}.$$

3–14

$$\mathbf{G} = \begin{pmatrix} 0.1818 & 0 & 0 \\ 0.0367 & 0.1645 & 0 \\ 0.0676 & 0.0390 & 0.1509 \end{pmatrix}, \quad \mathbf{F} = \begin{pmatrix} 5.50 & -1.23 & -2.15 \\ 0 & 6.08 & -1.57 \\ 0 & 0 & 6.63 \end{pmatrix}.$$

3–15 From Eq. (3–32), $\mathbf{I} = \mathbf{G}\overline{\mathbf{G}}$, and the results follow.

3–17 $A^{ii} = g'_{jk}\, A'^{jk}$, $B^{i}{}_{i} = B'^{i}{}_{i}$, $C_{ii} = g'^{jk}\, C'_{jk}$, where the unprimed tensor components are in the cartesian system and the primed components are in the general system.

3–18 The quantity given is proportional to the trace of $\mathbf{A}^{-1}$, which is invariant under an orthogonal transformation.

3–19 For example, for the symmetric case, $C'_{ij} = G_i^k\, G_j^l\, C_{kl} = G_j^l\, G_i^k\, C_{lk} = C'_{ji}$, and for the antisymmetric case $C'_{ij} = G_i^k\, G_j^l\, C_{kl} = -G_j^l\, G_i^k\, C_{lk} = -C'_{ji}$.

3–20

h_1	h_2	h_3	h'_1	h'_2	h'_3	h''_1	h''_2	h''_3
1	3	2	2	−1	−2	3	1	−1
2	4	0	3	−1	0	4	−1	2
3	−1	−4	1	2	4	−1	−2	7
1	−1	0	0	1	0	−1	1	1
2	0	−1	1	1	1	0	0	3

3–22 $g^{i}{}_{j} = \delta^{i}{}_{j}$.

3–24

$$\{\epsilon^{1}{}_{ij}\} = \begin{pmatrix} 0 & g^{13}V & -g^{12}V \\ -g^{13}V & 0 & g^{11}V \\ g^{12}V & -g^{11}V & 0 \end{pmatrix}, \quad \{\epsilon^{2}{}_{ij}\} = \begin{pmatrix} 0 & g^{23}V & -g^{22}V \\ -g^{23}V & 0 & g^{21}V \\ g^{22}V & -g^{21}V & 0 \end{pmatrix},$$

$$\{\epsilon^{3}{}_{ij}\} = \begin{pmatrix} 0 & g^{33}V & -g^{32}V \\ -g^{33}V & 0 & g^{31}V \\ g^{32}V & -g^{31}V & 0 \end{pmatrix}.$$

3–25 $x'^1 = -0.1400$, $x'^2 = 0.5533$, $x'^3 = 0.1533$. If $X^{ij} = x^i x^j$, then $\mathbf{X}' = \mathbf{F}\mathbf{X}\overline{\mathbf{F}}$.

3–27 (a) [14, −32, −13].

(b) $\mathbf{u}' = [3, -5, 2]$, $\mathbf{v}' = [1, -6, -4]$, $\mathbf{u}' \wedge \mathbf{v}' = -[32, 14, -13]$; the negative sign in front of the brackets takes into account that the new axes are left-handed. $\mathbf{F}\mathbf{u} \wedge \mathbf{v} = [-32, -14, 13]$.

3–28 (a) [17.08, −1.30, 1.92].

(b) $\mathbf{u}' = [0.167, -0.533, 0.033]$, $\mathbf{v}' = [0.600, -0.300, 1.400]$, $\mathbf{u}' \wedge \mathbf{v}' = -[-5.90, -7.20, 5.28]$; the negative sign in front of the brackets takes into account the change in handedness of the axes (G is negative). $\mathbf{F}\mathbf{u} \wedge \mathbf{v} = [5.90, 7.20, -5.28]$.

3–29

$$\beta' = \begin{pmatrix} 0.2505 & -0.0305 & 0.0542 \\ -0.0305 & 0.0224 & -0.0073 \\ 0.0542 & -0.0073 & 0.0356 \end{pmatrix}.$$

3–30 $\beta'^{11} = 0.777$, $\beta'^{12} = 0.066$, $\beta'^{13} = -0.057$, $\beta'^{22} = 0.773$, $\beta'^{23} = -0.240$, $\beta'^{33} = 0.858$.

3–31 $\beta^{11} = 0.0194$, $\beta^{22} = 0.0098$, $\beta^{33} = 0.0280$, $\beta^{12} = \beta^{23} = 0$, $\beta^{13} = 0.0076$.

3–32 (a)
$$W'_{IJ} = F^I{}_i F^J{}_j W_{ij} a^i a^j \,/\, (F^I{}_l F^I{}_m F^J{}_n F^J{}_o g^{lm} g^{no})^{1/2}.$$
There is no summation in this formula over I and J. (b) **W** is not a tensor.

3–33 [0.0635, 0.0863, 0.0283], [0.1030, −0.0500, 0.0615].

3–34
$$\mathbf{F} = \begin{pmatrix} -4.69 & 0.65 & 8.38 \\ 3.54 & 8.63 & 1.11 \\ 5.45 & -5.00 & 3.06 \end{pmatrix}, \qquad \mathbf{F}\beta\overline{\mathbf{F}} = \begin{pmatrix} 0.482 & 0 & 0 \\ 0 & 0.350 & 0 \\ 0 & 0 & 0.148 \end{pmatrix}$$

3–35 The eigenvalues are all equal to one, and all vectors are eigenvectors. This follows, since the metric tensor converts any vector into itself.

3–36 $\lambda = 0.5712$, [0.0248, 0.0515, 0.0801], 0.170 Å;

$\lambda = 0.7540$, [0.0770, −0.0030,0.0322], 0.195 Å;

$\lambda = 1.0835$, [−0.0174, −0.0239, 0.0930], 0.234 Å.

3–37
$$\mathbf{G} = \begin{pmatrix} 0.0249 & 0.0515 & 0.0801 \\ 0.0770 & -0.0029 & 0.0322 \\ -0.0174 & -0.0239 & 0.0932 \end{pmatrix}.$$

$\mathbf{G}\mathbf{g}\overline{\mathbf{G}} = \mathbf{I}.$

3–38 0.483, [0.0411, 0.0268, 0.1512], 0.156 Å;

0.448, [−0.0139, −0.0866, 0.0587], 0.151 Å;

0.677, [−0.1097, 0.0210, 0.0492], 0.185 Å.

3–39 2.409.

3–40 3.21 Å^2.

3–41 2.14 Å^2.

3–42 18.3°.

3–44 The result follows from Exercise 3–19; note that it does not apply to mixed tensors.

3–45 $\beta^{kk}/(a^k)^2$ (no summation).

3–46 (a) 0.651; (b) 0.463; (c) 0.496; (d) 0.677 Å^2.

3–48
$$A'^i{}_l = \epsilon'_{klm} r'^m g'^{ki} = J\epsilon_{klm} F^m{}_n F^k{}_o F^i{}_p r^n g^{op}$$
$$= JPVe_{klm} F^m{}_n F^k{}_o F^i{}_p r^n g^{op}.$$

By Eq. (3–86),

$$
\begin{aligned}
A'^i{}_l &= PVG_k{}^j\, G_l{}^s\, G_m{}^t\, e_{jst}F^m{}_n\, F^k{}_o\, F^i{}_p\, r^n g^{op} \\
&= PV\delta^j{}_o\, G_l{}^s\, \delta^t{}_n\, F^i{}_p\, e_{jst} r^n g^{op} \\
&= \epsilon_{jst}F^i{}_p\, G_l{}^s\, r^t g^{jp} = F^i{}_p\, G_l{}^s\, A^p{}_s.
\end{aligned}
$$

3–54 $\mathbf{I} = g^{ij}\,\mathbf{a}_i\mathbf{a}_j = g_{ij}\mathbf{a}^i\mathbf{a}^j$.

3–55 $g^{ij}\phi_{ij} = g_{ij}\phi^{ij} = \phi^i{}_i = \phi_i{}^i$.

3–56 $\phi^{ij} = u^i v^j$.

3–57 The result follows from $\mathbf{r} \wedge \mathbf{s} = -\mathbf{s} \wedge \mathbf{r}$.

3–62 $|\Phi| = |\Phi_C| = 1/|\Phi^{-1}|$, and the result follows.

3–63 $0.00906\mathbf{a}_1\mathbf{a}_1 - 0.00049(\mathbf{a}_1\mathbf{a}_2 + \mathbf{a}_2\mathbf{a}_1) - 0.00102(\mathbf{a}_1\mathbf{a}_3 + \mathbf{a}_3\mathbf{a}_1) + 0.00401\mathbf{a}_2\mathbf{a}_2 + 0.00038(\mathbf{a}_2\mathbf{a}_3 + \mathbf{a}_3\mathbf{a}_2) + 0.01424\mathbf{a}_3\mathbf{a}_3$.

3–64 The components are given by Exercise 3–29.

4–1 $\mathbf{S}$ and $\mathbf{R}$ commute, and $\mathbf{St} + \mathbf{u} = \mathbf{Ru} + \mathbf{t}$.

4–4 $\{\mathbf{I}|\mathbf{a}_3\}$.

4–5 $R'^i{}_i = F^i{}_k\, R^k{}_l\, G_i{}^l = \delta_k{}^l\, R^k{}_l = R^k{}_k$.

4–6 $\{\mathbf{I}|\mathbf{Ft}\}$.

4–7 $\{\mathbf{FR\bar{G}}|\mathbf{0}\}$.

4–8 $\{\mathbf{R}|\mathbf{Rv} + \mathbf{t} - \mathbf{v}\}$, $\mathbf{R}' = \mathbf{R}$, $\mathbf{t}' = \mathbf{Rv} + \mathbf{t} - \mathbf{v}$.

4–9 (a) $\begin{pmatrix} 0 & -1 & 0 \\ 1 & 0 & 0 \\ 0 & 0 & 1 \end{pmatrix}$. (b) $\begin{pmatrix} 0 & 0 & 1 \\ 1 & 0 & 0 \\ 0 & 1 & 0 \end{pmatrix}$. (c) $\begin{pmatrix} 0 & -1 & 0 \\ 0 & 0 & 1 \\ -1 & 0 & 0 \end{pmatrix}$.

(d) $\begin{pmatrix} -\tfrac{1}{2} & -\sqrt{3}/2 & 0 \\ \sqrt{3}/2 & -\tfrac{1}{2} & 0 \\ 0 & 0 & 1 \end{pmatrix}$. (e) $\begin{pmatrix} -1 & 0 & 0 \\ 0 & 1 & 0 \\ 0 & 0 & 1 \end{pmatrix}$.

4–10 $m = \bar{2}$.

4–11 $\begin{pmatrix} -\tfrac{1}{2} & -\sqrt{3}/2 & 0 \\ \sqrt{3}/2 & -\tfrac{1}{2} & 0 \\ 0 & 0 & 1 \end{pmatrix}$, $\begin{pmatrix} -\tfrac{1}{2} & \sqrt{3}/2 & 0 \\ -\sqrt{3}/2 & -\tfrac{1}{2} & 0 \\ 0 & 0 & 1 \end{pmatrix}$, $\begin{pmatrix} 1 & 0 & 0 \\ 0 & 1 & 0 \\ 0 & 0 & 1 \end{pmatrix}$.

4–12 (a)

$$\mathbf{R}_x(90°)\mathbf{R}_z(120°) = \begin{pmatrix} -\tfrac{1}{2} & -\sqrt{3}/2 & 0 \\ 0 & 0 & -1 \\ \sqrt{3}/2 & -\tfrac{1}{2} & 0 \end{pmatrix}.$$

(b) Note that powers of this operator do not repeat; this cannot be an operator of a finite group. (See also Exercise 4–50.)

4–13 Proper: $2\cos\theta + 1$.
Improper Schoenflies: $2\cos\theta - 1$.
Improper Hermann-Mauguin: $-2\cos\theta - 1$.

4–16 Besides the obvious relationships, such as $s^i s_i = 1$ and $s^i t_i = 0$, use the other consequences of the orthonormality of $\mathbf{s}$, $\mathbf{t}$, and $\mathbf{u}$, such as $s^1 s_1 + t^1 t_1 + u^1 u_1 = 1$, $s_1 s^2 + t_1 t^2 + u_1 u^2 = 0$, and the component relationships coming from $\mathbf{s} \wedge \mathbf{t} = \mathbf{u}$, etc.

4–18 (b) $$\begin{pmatrix} 2u^1u_1 - 1 & 2u^1u_2 & 2u^1u_3 \\ 2u^2u_1 & 2u^2u_2 - 1 & 2u^2u_3 \\ 2u^3u_1 & 2u^3u_2 & 2u^3u_3 - 1 \end{pmatrix}.$$

(c) $$\begin{pmatrix} 1 - 2u^1u_1 & -2u^1u_2 & -2u^1u_3 \\ -2u^2u_1 & 1 - 2u^2u_2 & -2u^2u_3 \\ -2u^3u_1 & -2u^3u_2 & 1 - 2u^3u_3 \end{pmatrix}.$$

(d) $\bar{\mathbf{1}} = -\mathbf{I}$.

4–19 $$\begin{pmatrix} \cos\theta + \cot\beta\sin\theta & 0 & a_3\sin\theta/(a_1\sin\beta) \\ 0 & 1 & 0 \\ -a_1\sin\theta/(a_3\sin\beta) & 0 & \cos\theta - \cot\beta\sin\theta \end{pmatrix}.$$

To be a symmetry operator in a monoclinic crystal, θ must be 0° or 180°.

4–20 $$(\mathbf{3})^2 = \begin{pmatrix} -1 & 1 & 0 \\ -1 & 0 & 0 \\ 0 & 0 & 1 \end{pmatrix}.$$

4–21 Let the given matrix be $\mathbf{G}^{-1}$. Applying Eq. (4–18) to Eq. (4–32) gives

$$\begin{pmatrix} 0 & 0 & 1 \\ 1 & 0 & 0 \\ 0 & 1 & 0 \end{pmatrix}.$$

4–22
$$\begin{pmatrix} (1 + 2\cos\theta)/3 & (1 - \cos\theta)/3 - \sin\theta/\sqrt{3} & (1 - \cos\theta)/3 + \sin\theta/\sqrt{3} \\ (1 - \cos\theta)/3 + \sin\theta/\sqrt{3} & (1 + 2\cos\theta)/3 & (1 - \cos\theta)/3 - \sin\theta/\sqrt{3} \\ (1 - \cos\theta)/3 - \sin\theta/\sqrt{3} & (1 - \cos\theta)/3 + \sin\theta/\sqrt{3} & (1 + 2\cos\theta)/3 \end{pmatrix}.$$

4–23 $$\begin{pmatrix} -1 & 1 & 0 \\ 0 & 1 & 0 \\ 0 & 0 & 1 \end{pmatrix}.$$

4–24 (a) $$\begin{pmatrix} 1 & 0 & 0 \\ 0 & \cos\theta & -a_3\sin\theta/a_2 \\ 0 & a_2\sin\theta/a_3 & \cos\theta \end{pmatrix}.$$ (b) $$\begin{pmatrix} \cos\theta & -\sin\theta & 0 \\ \sin\theta & \cos\theta & 0 \\ 0 & 0 & 1 \end{pmatrix}.$$

(c) $$\begin{pmatrix} \cos\theta + \sin\theta/\sqrt{3} & -2\sin\theta/\sqrt{3} & 0 \\ 2\sin\theta/\sqrt{3} & \cos\theta - \sin\theta/\sqrt{3} & 0 \\ 0 & 0 & 1 \end{pmatrix}.$$

(d) $$\begin{pmatrix} 1 & 0 & 0 \\ (1 - \cos\theta)/2 & \cos\theta & -a_3\sin\theta/a_1 \\ -a_1\sin\theta/(2a_3) & a_1\sin\theta/a_3 & \cos\theta \end{pmatrix}.$$

(e) Same as answer to Exercise 4–22.

(f) $\begin{pmatrix} (1 + \cos\theta)/2 & (1 - \cos\theta)/2 & \sin\theta/\sqrt{2} \\ (1 - \cos\theta)/2 & (1 + \cos\theta)/2 & -\sin\theta/\sqrt{2} \\ -\sin\theta/\sqrt{2} & \sin\theta/\sqrt{2} & \cos\theta \end{pmatrix}$.

4–25 (a) 0°, 180°; (b) 0°, 180°, 90°; (c) 0°, 60°, 120°, 180°; (d) 0°, 180°; (e) 0°, 60° (cell must be nonprimitive), 120°, 180° (nonprimitive cell); (f) 0°, 180°.

4–28 (a) $\pm[x, y, z; x - y, x, z; \bar{y}, x - y, z; \bar{x}, \bar{y}, z; y - x, \bar{x}, z; y, y - x, z]$.
(b) $[x, y, z; 2/3 + x - y, 1/3 + x, z; \bar{y}, x - y, z; 2/3 - x, 1/3 - y, z; y - x, \bar{x}, z; 2/3 + y, 1/3 + y - x, z; 2/3 - x, 1/3 - y, \bar{z}; y - x, \bar{x}, \bar{z}; 2/3 + y, 1/3 - x + y, \bar{z}; x, y, \bar{z}; 2/3 + x - y, 1/3 + x, \bar{z}; \bar{y}, x - y, \bar{z}]$.

4–29 (a) $\pm[x, y, z; y - x, \bar{x}, z; \bar{y}, x - y, z; \bar{y}, \bar{x}, z; x, x - y, z; y - x, y, z]$.
(b) $\pm[x, y, z; 1/2 - y, 1/2 + x - y, z; y - x, 1/2 - x, z; 1/2 - y, 1/2 - x, z; x, 1/2 + x - y, z; y - x, y, z]$.

4–30 $\pm(h, k, l; h + k, \bar{h}, l; k, -h - k, l; \bar{h}, \bar{k}, l; -h - k, h, l; \bar{k}, h + k, l)$.

4–32 $[x, y, z; y, \bar{x}, \bar{z}; \bar{x}, \bar{y}, z; \bar{y}, x, \bar{z}; x, \bar{y}, \bar{z}; y, x, z; \bar{x}, y, \bar{z}; \bar{y}, \bar{x}, z]$
$\pm(h, k, l; \bar{k}, h, \bar{l}; \bar{h}, \bar{k}, l; k, \bar{h}, \bar{l}; h, \bar{k}, \bar{l}; k, h, l; \bar{h}, k, \bar{l}; \bar{k}, \bar{h}, l)$.

4–33 $g_{22} = g_{11}, g_{12} = -g_{11}/2, g_{13} = g_{23} = 0$.

4–34 (a) $\beta'^{11} = \beta^{22}$, $\beta'^{12} = -\beta^{12} + \beta^{22}$, $\beta'^{13} = -\beta^{23}$, $\beta'^{22} = \beta^{11} - 2\beta^{12} + \beta^{22}$, $\beta'^{23} = \beta^{13} - \beta^{23}$, $\beta'^{33} = \beta^{33}$.
(b) $\beta^{22} = \beta^{11}, \beta^{12} = \beta^{11}/2, \beta^{13} = \beta^{23} = 0$.

4–35 (a) $\beta'^{11} = \beta^{11} - 2\beta^{12} + \beta^{22}$, $\beta'^{12} = -\beta^{12} + \beta^{22}$, $\beta'^{13} = -\beta^{13} + \beta^{23}$, $\beta'^{22} = \beta^{22}$, $\beta'^{23} = \beta^{23}$, $\beta'^{33} = \beta^{33}$.
(b) $\beta^{12} = \beta^{22}/2, \beta^{23} = 2\beta^{13}$.

4–36 (a) $\beta'^{11} = \beta^{22}$, $\beta'^{12} = -\beta^{12}$, $\beta'^{13} = -\beta^{23}$, $\beta'^{22} = \beta^{11}$, $\beta'^{23} = \beta^{13}$, $\beta'^{33} = \beta^{33}$.
(b) $\beta^{22} = \beta^{11}, \beta^{12} = \beta^{13} = \beta^{23} = 0$.

4–37 (a) $\beta'^{11} = \beta^{11}$, $\beta'^{12} = -\beta^{12}$, $\beta'^{13} = -\beta^{13}$, $\beta'^{22} = \beta^{22}$, $\beta'^{23} = \beta^{23}$, $\beta'^{33} = \beta^{33}$.
(b) $\beta^{12} = \beta^{13} = 0$.

4–38 (a) $\beta'^{11} = \beta^{22}$, $\beta'^{12} = \beta^{12}$, $\beta'^{13} = -\beta^{23}$, $\beta'^{22} = \beta^{11}$, $\beta'^{23} = -\beta^{13}$, $\beta'^{33} = \beta^{33}$.
(b) $\beta^{11} = \beta^{22}, \beta^{23} = -\beta^{13}$.

4–39 $\beta'^{11} = \beta^{33}$, $\beta'^{12} = \beta^{13}$, $\beta'^{13} = \beta^{23}$, $\beta'^{22} = \beta^{11}$, $\beta'^{23} = \beta^{12}$, $\beta'^{33} = \beta^{22}$; $\beta^{11} = \beta^{22} = \beta^{33}, \beta^{12} = \beta^{13} = \beta^{23}$.

4–40 $\beta^{12} = \beta^{23} = 0$.

4–41 $\beta^{12} = \beta^{13} = \beta^{23} = 0$.

4–42 $\mathbf{g}^* = \mathbf{R}^{-1}\mathbf{g}^*\bar{\mathbf{R}}^{-1} = \mathbf{R}\mathbf{g}^*\bar{\mathbf{R}}$.

4–43 The result is obviously true in a cartesian system. In a general system related to the cartesian system by Eq. (4–18), $|\mathbf{R}'| = |\mathbf{F}|\ |\mathbf{R}|\ |\overline{\mathbf{G}}| = |\mathbf{R}|$, since $|\mathbf{F}| = 1/|\overline{\mathbf{G}}|$.

4–45 (a) $\mathbf{w} = [a_3/\sqrt{3},\ -16a_3/\sqrt{3},\ -13\sqrt{3}(a_1)^2/a_3]$.
(b) $\mathbf{s} = [4,\ 1,\ 1]$, $\mathbf{t} = (-2,\ -7,\ -2]$.
(c) $\mathbf{r} = [16a_3/\sqrt{3},\ 17a_3/\sqrt{3},\ -13\sqrt{3}\,(a_1)^2/a_3]$; $RR'_n\ w^n = (-1)[-16a_3/\sqrt{3},$ $-17a_3/\sqrt{3},\ 13\sqrt{3}(a_1)^2/a_3]$.

4–47 The reflection operator has eigenvalues 1, 1, and -1. The eigenvectors associated with the doubly degenerate eigenvalues are any two orthonormal vectors of the form $[x,\ 0,\ z]$. The eigenvector associated with eigenvalue -1 is [0, 1, 0]. The identity operator has eigenvalues 1, 1, and 1, and any three orthonormal vectors are eigenvectors.

4–48 $\lambda = 1$, $\mathbf{v} = [0,\ 0,\ 1]$;
$\lambda = e^{i\pi/2}$, $\mathbf{v} = [1/\sqrt{2},\ -i/\sqrt{2},\ 0]$;
$\lambda = e^{-i\pi/2}$, $\mathbf{v} = [1/\sqrt{2},\ i/\sqrt{2},\ 0]$;
where $i = \sqrt{-1}$.

4–49 $\lambda = 1$, $\mathbf{v} = [1/\sqrt{3},\ 1/\sqrt{3},\ 1/\sqrt{3}]$;
$\lambda = e^{2\pi i/3}$, $\mathbf{v} = [1/\sqrt{3},\ e^{2\pi i/3}/\sqrt{3},\ e^{-2\pi i/3}/\sqrt{3}]$;
$\lambda = e^{-2\pi i/3}$, $\mathbf{v} = [1/\sqrt{3},\ e^{-2\pi i/3}/\sqrt{3},\ e^{2\pi i/3}/\sqrt{3}]$.

4–50 $\lambda = 1$, $\mathbf{v} = [1/\sqrt{7},\ \sqrt{3}/\sqrt{7},\ \sqrt{3}/\sqrt{7}]$; vector $\mathbf{v}$ is the rotation axis.
$\lambda = (-3 \pm \sqrt{7}i)/4 = e^{\pm i\theta}$, where $\theta = \cos^{-1}(-3/4) = 138.59°$.
This angle is not commensurate with 360°, so a finite number of applications of the operator will not produce the identity. (See also Exercise 4–12.)

4–51 A rotation of $\cos^{-1}(-7/8) = 151.04°$. If the combination is $\mathbf{R}_z\mathbf{R}_x$, the rotation axis is $[1/\sqrt{5},\ \sqrt{3}/\sqrt{5},\ 1/\sqrt{5}]$. The rotation is not commensurate with 360°, so it is not a possible symmetry operation in a finite group.

4–52 ±120° rotation about [001]; hexagonal or trigonal system.

4–53 180° rotation about [100]. The eigenvectors corresponding to the doubly degenerate eigenvalues -1 are normal to [100].

4–54 Mirror plane normal to [010].

4–55 180° rotation about [210]. Other eigenvectors normal to [210].

4–56 Mirror plane $= \overline{2}$ normal to $[1,\ -1,\ 0]$.

4–57 $\overline{4}$ rotation about [001].

4–58 180° rotation about [110].

5–1 (a) $\sigma^{11} = 9.17 \times 10^3$, $\sigma^{33} = 7.25 \times 10^3$ ohm^{-1} cm^{-1}
(b) $\sigma^{11} = 594$, $\sigma^{12} = 297$, $\sigma^{33} = 51.7$ ohm^{-1}cm^{-1}Å^{-2}
(c) $138 \times 10^{-6} \times 3.00/(\pi \times 0.500 \times 0.500) = 5.27 \times 10^{-4}$ ohm.
(d) In the cartesian system, $\mathbf{u} = [0.479,\ 0.277,\ 0.833]$, $\sigma(\mathbf{u}) = 7.84 \times 10^3$, $R = 4.87 \times 10^{-4}$ ohm. Note that basing the calculation upon $\rho(\mathbf{u})$, which would lead to $R = 4.93 \times 10^{-4}$ ohm, would not be consistent with the experimental configuration.

5–2 7.84×10^3 ohm^{-1} cm^{-1} for all three directions.

5–3 $\chi^{ii}/3$.

5-4 (a) $\chi^{11} = 0.155$, $\chi^{12} = 0.173$, $\chi^{13} = 0.397$, $\chi^{22} = -0.0776$, $\chi^{23} = 0.212$, $\chi^{33} = 0.315$.

(b) $\lambda = 0.732$, $\mathbf{v} = [0.600, 0.320, 0.733]$;

$\lambda = -0.170$, doubly degenerate, $\mathbf{v}$ any linear combination of $[0.202, -0.947, 0.248]$ and $[-0.775, 0, 0.632]$.

(c) $\mathbf{a}_3$ must be parallel to $[0.600, 0.320, 0.733]$. A suitable transformation is

$$\begin{pmatrix} \mathbf{a}_1 \\ \mathbf{a}_2 \\ \mathbf{a}_3 \end{pmatrix} = \begin{pmatrix} 0.202a_1 & -0.947a_1 & 0.248a_1 \\ -0.775a_1 & 0 & 0.632a_1 \\ 0.600a_3 & 0.320a_3 & 0.733a_3 \end{pmatrix} \begin{pmatrix} \mathbf{e}_1 \\ \mathbf{e}_2 \\ \mathbf{e}_3 \end{pmatrix}.$$

(d) When referred to principal axes, the diagonalized tensor has values -0.170, -0.170, 0.732, The magnetic susceptibility will be 0 along directions making angle $\tan^{-1}\sqrt{(0.732/0.170)} = 64.3°$ with $\mathbf{a}_3$.

(e) By Exercise 5-3, polycrystalline zircon would be paramagnetic with susceptibility 0.131.

5-5 $A^{11} = A'^{11}\cos^2\theta + A'^{22}\sin^2\theta$; $A^{22} = A'^{11}\sin^2\theta + A'^{22}\cos^2\theta$; $A^{12} = (A'^{22} - A'^{11})\sin\theta\cos\theta$.

5-6 (a) Conductivity = 0.0649 watt/deg cm along [211], 0.0580 watt/deg cm along [101].

(b) 0.35 watt.

5-7 (a) 4.9×10^{-4} cal/cm-deg-sec.

(b) 3.8×10^{-4} cal/cm-deg-sec.

5-8 A hyperboloid of revolution of two sheets.

5-9 Representation quadric is an oblate ellipsoid (or spheroid) with semi-axis of length 43 parallel to the grain, of length 58 across the grain. Magnitude ellipsoid is a prolate ellipsoid with semi-axis lengths 5.5×10^{-4} parallel to the grain, 3.0×10^{-4} across the grain.

5-10 $\sigma^{33} = 7.50/4.00$ newtons/cm²; all other components are zero.

5-11 $\sigma^{33} = (60.0)(970)/5.00$ dynes/cm²; all other components are zero.

5-12 $\pm\sigma^{12}$.

5-13 $\sigma' = -p\mathbf{g}^*$.

5-14 (a) Nonvanishing components are $\epsilon_{12} = \epsilon_{21} = (\sin\alpha)/2$, $\epsilon_{22} = (\sin^2\alpha)/2$.

(b)
$$\mathbf{G} = \begin{pmatrix} 1 & 0 & 0 \\ \sin\alpha & 1 & 0 \\ 0 & 0 & 1 \end{pmatrix}, \qquad \mathbf{g}' = \begin{pmatrix} 1 & \sin\alpha & 0 \\ \sin\alpha & 1+\sin^2\alpha & 0 \\ 0 & 0 & 1 \end{pmatrix}.$$

5-15 (a)
$$\epsilon = \begin{pmatrix} -10.12\sin\phi + 28.12\sin^2\phi & 0 & 46.12\sin\phi - 5.06\sin^2\phi \\ 0 & 0 & 0 \\ 46.12\sin\phi - 5.06\sin^2\phi & 0 & -10.12\sin\phi + 18.00\sin^2\phi \end{pmatrix}.$$

(b)
$$\begin{pmatrix} -10.12\phi & 0 & 46.12\phi \\ 0 & 0 & 0 \\ 46.12\phi & 0 & -10.12\phi \end{pmatrix}.$$

(c) $u^1 = x^3 \sin\phi$, $u^2 = 0$, $u^3 = x^1 \sin\phi$; nonvanishing elements of $\mathbf{U}$ are $U_{11} = U_{33} = -10.12 \sin\phi$, $U_{13} = 36.00 \sin\phi$, $U_{31} = 56.25 \sin\phi$.

(d) $$\begin{pmatrix} -0.281\phi & 0 & 0.987\phi \\ 0 & 0 & 0 \\ 0.987\phi & 0 & 0.281\phi \end{pmatrix}.$$

(e) $\pm 1.03\phi$.

5–16 $\lambda = 0$, $\mathbf{v} = [001]$;
$\lambda = \epsilon_{12}$, $\mathbf{v} = [1/\sqrt{2}, 1/\sqrt{2}, 0]$;
$\lambda = -\epsilon_{12}$, $\mathbf{v} = [1/\sqrt{2}, -1/\sqrt{2}, 0]$.

5–17 (a) From Eq. (5–31), $\mathbf{g}' = 2\epsilon + \mathbf{g} = \mathbf{g}(2\mathbf{g}^*\epsilon + \mathbf{I})$; $g' = g\,|2\mathbf{g}^*\epsilon + \mathbf{I}|$. If the components of ϵ are small, $|2\mathbf{g}^*\epsilon + \mathbf{I}| = \text{trace}(2\mathbf{g}^*\epsilon) + 1$. But $\text{trace}(\mathbf{g}^*\epsilon) = g^{ij}\epsilon_{ij}$, and $g' = V'^2$, $g = V^2$, so $V'^2 = V^2(2g^{ij}\epsilon_{ij} + 1)$. Taking the square root and using $(1 + x)^{1/2} \simeq 1 + \frac{1}{2}x$ *for small* x, $\Delta V = Vg^{ij}\epsilon_{ij}$.
(b) $V' = 718.23$, $V = 717.22$, $g^{ij}\epsilon_{ij} = 0.00141$, $\Delta V = 1.01$.

5–19 (a) $\alpha_{11} = \alpha_{22} = 1.8413 \times 10^{-4} a_1$, $\alpha_{33} = 1.6588 \times 10^{-4} a_3$.
(b) $\alpha'_{11} = \alpha'_{22} = 1.8413 \times 10^{-4}/a_1$, $\alpha'_{33} = 1.6588 \times 10^{-4}/a_3$.
(c) [110]: $\alpha = 4.3356 \times 10^{-5}\,K^{-1}$;
[001]: $\alpha = 2.0783 \times 10^{-5}\,K^{-1}$;
[111]: $\alpha = 3.5832 \times 10^{-5}\,K^{-1}$.

5–20 A cartesian system may be defined in which P is at (0, 0), Q at $(\sqrt{3}x_1/2, x_1/2)$, R at $(-\sqrt{3}x_2/2, x_2/2)$, S at $(0, -x_3)$, and C at $[(x_1 - x_2)/\sqrt{3}, x_1 + x_2 - 2/3]$.

5–22 (a) $i_1 = 48.2 \times 10^{-6}$, $i_2 = -2.3 \times 10^{-10}$, $A = 0.76$.
(b) Point P is outside the triangle, to the lower right of vertex 2.
(c) $48.2 \times 10^{-6}\,K^{-1}$.

5–26 $c^{16} = c^{11}/2$, $c^{22} = c^{11}$, $c^{23} = c^{13}$, $c^{24} = -c^{15}$, $c^{25} = c^{14} - c^{15}$, $c^{26} = c^{11}/2$, $c^{34} = 0$, $c^{35} = 0$, $c^{36} = c^{13}/2$, $c^{45} = c^{44}/2$, $c^{55} = c^{44}$, $c^{46} = c^{14} - c^{15}$, $c^{56} = c^{14}$, $c^{66} = (3c^{11} - 2c^{12})/4$.

5–27 $c^{11} = c^{22}$, $c^{13} = c^{23}$, $c^{14} = -c^{24} = c^{56}$, $c^{44} = c^{55}$, $c^{15} = -c^{25} = -c^{46}$, $c^{66} = (c^{11} - c^{12})/2$. All others 0.

5–28 $c'^{11} = 9(a_1)^4 c^{11}/16$, $c'^{12} = 3(a_1)^4(4c^{12} - c^{11})/16$, $c'^{13} = 3(a_1a_3)^2 c^{13}/4$, $c'^{14} = 3\sqrt{3}(a_1)^3 a_3 c^{15}/8$, $c'^{15} = 3(a_1)^3 a_3(c^{15} - 2c^{14})/8$, $c'^{16} = 0$, $c'^{33} = (a_3)^4 c^{33}$, $c'^{44} = 3(a_1a_3)^2 c^{44}/4$.

5–29 $\epsilon_{ij} = s_{ijkl}\sigma^{kl} = s_{ijkl}c^{klmn}\epsilon_{mn} = \delta_i{}^m\delta_j{}^n\epsilon_{mn}$.

5–30 (a) 243; (b) 63, of which 18 are zero.

6–2 (a) Rotate through angle $QOQ' = \theta$.
(b) Q' will miss the sphere if $h > 2/\lambda$, or $\sigma > 2$.

6–3 $\mathbf{X} = (h_1 - h_3 a_1 \cos\beta/a_3)\mathbf{a}^1 + (h_2 - h_3 a_2 \cos\alpha/a_3)\mathbf{a}^2$.

6–4 $\sin\phi = (h_1 a^1 + h_2 a^2 \cos\gamma^* + h_3 a^3 \cos\beta^*)/X$.

6–5 $z = r_F Z\lambda/[1 - (Z\lambda)^2]^{1/2}$;

$y = r_F \cos^{-1} [(1 - \lambda^2 h^2/2)/(1 - \lambda^2 Z^2)^{1/2}]$.

6-6 $z = DZ\lambda/(1 - \lambda^2 h^2/2)$;
$y = D(4\lambda^2 X^2 - \lambda^4 h^4)^{1/2}/(2 - \lambda^2 h^2)$.

6-7 $\mathbf{t} = (0.09524, 0, 0]$, $\mathbf{u} = [0.10248, 0, 0.05124] = (10.047, 0, -0.577)$, $\mathbf{t} \wedge \mathbf{u} = [0, -0.03364, 0] = (0, -2.511, 0)$, $\chi = 16.89°$, $\Delta T = 7.98°$, $T = 22.01°$, $B = -1.61°$.

6-8 $\chi = -35.75°$.

6-10 $Z = 0.375$, $h(203) = 0.502$, $\Upsilon(203) = 31.14°$, $h(053) = 0.807$, $\Upsilon(053) = 70.23°$; (203) will be 15.57 mm from center line, (053) will be 35.12 mm from center line. These spots will lie on lines 45 mm apart measured along $\mathbf{e}_3$.

6-11 $\Upsilon(203) = 54.50°$, at 27.25 mm from center line;
$\Upsilon(\bar{2}03) = 27.07°$, at 13.54 mm from center line;
$\Upsilon(033) = 38.86°$, at 19.43 mm from center line.
$\mathbf{X}(203) = 2.65\mathbf{a}^1$, $\mathbf{X}(\bar{2}03) = -1.35\mathbf{a}^1$, $\mathbf{X}(033) = 0.65\mathbf{a}^1 + 3\mathbf{a}^2$. The angle between $\mathbf{X}(203)$ and $\mathbf{X}(\bar{2}03)$ is 180°; lines of slope 2 through these spots are 90 mm apart measured along $\mathbf{e}_3$. The angle between $X(203)$ and $X(033)$ is 70.34°; these reflections are on lines 35.17 mm apart.

6-12 $1/L = \cos^2\nu \sin \Upsilon = \lambda X \cos \theta$.

6-13 $R = \sin \nu/\lambda$, $R_0 = \sin \mu/\lambda$.

6-14 (a) $\Delta\phi$ is the angle between $\mathbf{e}_2'$ and $\mathbf{e}_2$, so by Eq. (6-37) $\cos \Delta\phi = \sin \chi$. This may be deduced also as the complement of the angle between $\mathbf{e}_1'$ and $\mathbf{e}_2$.
(b) $\Delta\phi = 0°$ for $\omega = 0°$ or 180°; ϕ ranges from μ for $\omega = 90°$ to $-\mu$ for $\omega = 270°$.

6-17 $\xi = 0.814834$.

6-18 $\cos \eta = \xi$, $\eta = \pm 35.43°$, $\omega = -58.36°$ or 238.36°. With $\eta = +35.43°$:

$$\mathbf{s}_0 = -1.2186\mathbf{e}_1' - 0.5733\mathbf{e}_2' - 0.4078\mathbf{e}_3';$$
$$\mathbf{s} = -1.2186\mathbf{e}_1' + 0.5733\mathbf{e}_2' - 0.4078\mathbf{e}_3';$$
$$\frac{d\mathbf{s}}{d\omega} = 0.4314\mathbf{e}_2' - 0.6063\mathbf{e}_3';$$
$$(\mathbf{s} - \mathbf{s}_0) \cdot \frac{d\mathbf{s}}{d\omega} = 0.4946.$$

With $\eta = -35.43°$:

$$\mathbf{s}_0 = -1.2186\mathbf{e}_1' - 0.5733\mathbf{e}_2' + 0.4078\mathbf{e}_3';$$
$$\mathbf{s} = -1.2186\mathbf{e}_1' + 0.5733\mathbf{e}_2' + 0.4078\mathbf{e}_3';$$
$$\frac{d\mathbf{s}}{d\omega} = -0.4314\mathbf{e}_2' - 0.6063\mathbf{e}_3';$$
$$(\mathbf{s} - \mathbf{s}_0) \cdot \frac{d\mathbf{s}}{d\omega} = -0.4946.$$

$p = 0.7231$, $L = (1/0.4946 + 1/0.4946)/2$;
$1/Lp = 0.6840$.

6–19 From Eqs. (6–68) and (6–69, $1/Lp$ = 0.22518. The method used in Exercise 6–18 includes an extra factor of $2/(\lambda)^2$. $[2/(\lambda)^2](0.22518) = 0.8916$. The value relative to the zero-level maximum of Exercise 6–18 is 0.8916/0.6840 = 1.304.

6–20 $D = 5.50$, $\eta = +90.459°$, $\mathbf{v} - \mathbf{c} = 21.18\mathbf{e}_2' + 11.85\mathbf{e}_3'$;
$D = 5.50$, $\eta = -90.459°$, $\mathbf{v} - \mathbf{c} = 20.85\mathbf{e}_2' + 12.41\mathbf{e}_3'$;
$D = 6.00$, $\eta = \pm 90.459°$, $\mathbf{v} - \mathbf{c} = 20.78\mathbf{e}_2' + 12.00\mathbf{e}_3'$.

6–21
$h(200) = 0.2687$, $\phi = 99.62°$, $\chi = 14.30°$;
$h(211) = 0.3395$, $\phi = 122.63°$, $\chi = 33.55°$;
$h(132) = 0.4507$, $\phi = 168.19°$, $\chi = 37.72°$;
$h(032) = 0.4110$, $\phi = 189.62°$, $\chi = 36.16°$;
$h(0\bar{3}2) = 0.4110$, $\phi = 9.62°$, $\chi = 36.16°$;
$h(130) = 0.3580$, $\phi = 168.19°$, $\chi = 5.32°$;
$h(030) = 0.3319$, $\phi = 189.62°$, $\chi = 0.00°$;
$h(\bar{1}30) = 0.3580$, $\phi = 211.04°$, $\chi = -5.32°$.

6–22

$$\mathbf{U} = \begin{pmatrix} -0.12642 & -0.02134 & 0 \\ 0 & 0 & -0.08913 \\ 0.01381 & -0.08183 & 0 \end{pmatrix}.$$

(a) (i) $h(022) = 0.2436$, $\phi = 170.42°$, $\chi = 47.04°$;
(ii) $h(432) = 0.6017$, $\phi = 242.49°$, $\chi = 26.38°$;
(iii) $h(\bar{4}32) = 0.6017$, $\phi = 98.35°$, $\chi = 26.38°$.
(b) (i) $h(022) = 0.2436$, $\sin\chi > 1$, not observable;
(ii) $h(432) = 0.6017$, $\phi = 190.37°$, $\chi = 38.94°$;
$\phi = 114.61°$, $\chi = 141.06°$;
(iii) $h(\bar{4}32) = 0.6017$, $\phi = 46.23°$, $\chi = 38.94°$;
$\phi = 330.48°$, $\chi = 141.06°$;
(iv) $h(200) = 0.2564$, $\phi = 215.42°$, $\chi = 0.00°$;
$\phi = 125.42°$, $\chi = 180.00°$.

6–23 $h(4\bar{2}\bar{2}) = 0.2564$, $\chi = 0.00°$;

$$\mathbf{U} = \begin{pmatrix} -0.05333 & 0.00783 & -0.03566 \\ 0.01976 & 0.03700 & 0.01783 \\ 0 & 0 & -0.08915 \end{pmatrix};$$

$h(\bar{2}\bar{4}2) = 0.2436$, $\chi = 47.04°$, $\phi = 170.42°$;
$h(6\bar{8}\bar{1}) = 0.6017$, $\chi = 26.38°$, $\phi = 242.49°$;
$h(\overline{10}\,0\,7) = 0.6018$, $\chi = 26.38°$, $\phi = 98.35°$.

6–24 (a) $\chi = 46.10°$.

$$\mathbf{U} = \begin{pmatrix} 0 & 0 & -0.118309 \\ 0.099142 & 0.025861 & -0.059155 \\ 0.020160 & -0.077286 & 0 \end{pmatrix}.$$

(b) $\chi = 67.95°$, $\phi = 165.38°$.
(c) $\chi = 66.05°$, $\phi = 124.84°$.
(d) 90.00°.

6–26 From Eq. (6–98), maximum $\omega = 50°$; $\chi = 90°$, $\Delta\phi = -90°$.

6–27 0°, 180°.

6–28 $\chi = 67.16°$, $\Delta\phi = -43.15°$; or $\chi = 112.84°$, $\Delta\varphi = 223.15°$.

6–29 (a) $\mathbf{e}_1' = -0.9978\mathbf{e}_1 - 0.0319\mathbf{e}_2 + 0.0583\mathbf{e}_3$;
$\mathbf{e}_2' = 0.0656\mathbf{e}_1 - 0.6131\mathbf{e}_2 + 0.7872\mathbf{e}_3$;
$\mathbf{e}_3' = 0.0106\mathbf{e}_1 + 0.7893\mathbf{e}_2 + 0.6139\mathbf{e}_3$.

(b) $\delta = 24.23°$, $\chi = -52.13°$, $\phi_E = 184.23°$, $\omega_E = -0.77°$.
(c) Same as (a).

6–30 $\varkappa = 216.43°$, $\omega = 170.37°$, $\psi = 102.50°$.
$\varkappa = 23.99°$, $\omega = 9.63°$, $\psi = 3.31°$.

6–31 (210): $\phi_\varkappa = 128.20°$, $\omega_\varkappa = 5.57°$, $\varkappa = -17.26°$;
($\bar{1}32$): $\phi_\varkappa = 224.23°$, $\omega_\varkappa = 13.19°$, $\varkappa = -40.06°$.

7–2 $\pi(3a - b)b^2/3$.

7–3 $\pi hR^2/3$.

7–7 $\pi R(R^2 + h^2)^{1/2}$.

7–8 $\pi r^2/2$.

7–9 $2r^2\omega$.

7–10
$$\frac{\partial \mathbf{a}_1}{\partial r} = \mathbf{0},\ \frac{\partial \mathbf{a}_1}{\partial \theta} = \frac{\mathbf{a}_2}{r},\ \frac{\partial \mathbf{a}_1}{\partial \phi} = \frac{\mathbf{a}_3}{r};$$

$$\frac{\partial \mathbf{a}_2}{\partial r} = \frac{\mathbf{a}_2}{r},\ \frac{\partial \mathbf{a}_2}{\partial \theta} = -r\mathbf{a}_1,\ \frac{\partial \mathbf{a}_2}{\partial \phi} = \cot\theta\ \mathbf{a}_3;$$

$$\frac{\partial \mathbf{a}_3}{\partial r} = \frac{\mathbf{a}_3}{r},\ \frac{\partial \mathbf{a}_3}{\partial \theta} = \cot\theta\ \mathbf{a}_3,\ \frac{\partial \mathbf{a}_3}{\partial \phi} = -r\sin^2\theta\ \mathbf{a}_1 - \sin\theta\cos\theta\ \mathbf{a}_2.$$

7–11 Differentiate $\mathbf{a}_j = (\partial l^i/\partial x^j)\mathbf{e}_i$ with respect to x^k (l^i is a cartesian coordinate). Exchanging the order of differentiation and replacing $(\partial l^i/\partial x^k)\mathbf{e}_i$ by $\mathbf{a}_k$ gives the result.

7–12 (a)
$$\mathbf{F} = \begin{pmatrix} \cos\theta\cos\chi & \sin\theta\cos\chi & \sin\chi \\ \dfrac{-\sin\theta}{b + r\cos\chi} & \dfrac{\cos\theta}{b + r\cos\chi} & 0 \\ \dfrac{-\cos\theta\sin\chi}{r} & \dfrac{-\sin\theta\sin\chi}{r} & \dfrac{\cos\chi}{r} \end{pmatrix}.$$

$$\mathbf{G} = \begin{pmatrix} \cos\theta\cos\chi & \sin\theta\cos\chi & \sin\chi \\ -\sin\theta(b + r\cos\chi) & \cos\theta(b + r\cos\chi) & 0 \\ -r\cos\theta\sin\chi & -r\sin\theta\sin\chi & r\cos\chi \end{pmatrix}.$$

(b) $J = r\,(b + r\cos\chi)$.

(c) $\mathbf{a}_1 = \cos\theta\cos\chi\mathbf{e}_1 + \sin\theta\cos\chi\mathbf{e}_2 + \sin\chi\,\mathbf{e}_3$;
$\mathbf{a}_2 = -\sin\theta\,(b + r\cos\chi)\mathbf{e}_1 + \cos\theta(b + r\cos\chi)\mathbf{e}_2$;
$\mathbf{a}_3 = -r\cos\theta\sin\chi\,\mathbf{e}_1 - r\sin\theta\sin\chi\mathbf{e}_2 + r\cos\chi\,\mathbf{e}_3$.

(d)
$$\mathbf{g} = \begin{pmatrix} 1 & 0 & 0 \\ 0 & (b + r\cos\chi)^2 & 0 \\ 0 & 0 & (r)^2 \end{pmatrix}.$$

(e) A torus with major radius b, minor radius a; $V = 2\pi^2a^2b$.
(f) $A = 4\pi^2ab$.

7–13
$$\mathbf{g} = \begin{pmatrix} \dfrac{\eta + \xi}{4\xi} & 0 & 0 \\ 0 & \dfrac{\eta + \xi}{4\eta} & 0 \\ 0 & 0 & \xi\eta \end{pmatrix}.$$

$J = (\eta + \xi)/4,\ dV = (\eta + \xi)\,d\eta\,d\xi\,d\phi/4.$

7–14 (a)
$$\mathbf{g} = \begin{pmatrix} 1 & 0 & 0 \\ 0 & (r)^2 + (s)^2 & s\theta \\ 0 & s\theta & (\theta)^2 \end{pmatrix}.$$

(b) $dV = r\,\theta\,dr\,ds\,d\theta$.

7–15 (a) $\begin{Bmatrix} 1 \\ 2\ 2 \end{Bmatrix} = -r$; $\begin{Bmatrix} 1 \\ 3\ 3 \end{Bmatrix} = -r\sin^2\theta$;

$$\begin{Bmatrix} 2 \\ 1\ 2 \end{Bmatrix} = \begin{Bmatrix} 2 \\ 2\ 1 \end{Bmatrix} = \begin{Bmatrix} 3 \\ 1\ 3 \end{Bmatrix} = \begin{Bmatrix} 3 \\ 3\ 1 \end{Bmatrix} = \frac{1}{r};$$

$$\begin{Bmatrix} 3 \\ 2\ 3 \end{Bmatrix} = \begin{Bmatrix} 3 \\ 3\ 2 \end{Bmatrix} = \cot\theta;\ \begin{Bmatrix} 2 \\ 3\ 3 \end{Bmatrix} = -\sin\theta\cos\theta.$$

All other $\begin{Bmatrix} i \\ j\ k \end{Bmatrix}$ vanish.
$[12,2] = [21,2] = r$; $[13,3] = [31,3] = r\sin^2\theta$;
$[22,1] = -r$; $[23,3] = [32,3] = (r)^2\sin\theta\cos\theta$;
$[33,1] = -r\sin^2\theta$; $[33,2] = -(r)^2\sin\theta\cos\theta$.
All other $[ij,k]$ vanish.

(b) $\left\{ {1 \atop 2\ 2} \right\} = -\cos\chi(b + r\cos\chi)$; $\left\{ {1 \atop 3\ 3} \right\} = -r$;

$$\left\{ {2 \atop 1\ 2} \right\} = \left\{ {2 \atop 2\ 1} \right\} = \cos\chi/(b + r\cos\chi);$$

$$\left\{ {2 \atop 2\ 3} \right\} = \left\{ {2 \atop 3\ 2} \right\} = -r\sin\chi/(b + r\cos\chi);$$

$$\left\{ {3 \atop 1\ 3} \right\} = \left\{ {3 \atop 3\ 1} \right\} = \frac{1}{r};\ \left\{ {3 \atop 2\ 2} \right\} = \frac{\sin\chi\,(b + r\cos\chi)}{r}.$$

All other $\left\{ {i \atop j\,k} \right\}$ vanish.

$[12, 2] = [21, 2] = -[22, 1] = \cos\chi\,(b + r\cos\chi)$;
$[13, 3] = [31, 3] = -[33, 1] = r$;
$[22, 3] = -[23,2] = -[32,2] = r\sin\chi\,(b + r\cos\chi)$.

All other $[ij,k]$ vanish.

7-16 $\operatorname{div} \mathbf{v} = \dfrac{\partial v^1}{\partial x^1} + \dfrac{\partial v^2}{\partial x^2} + \dfrac{\partial v^3}{\partial x^3}$.

7-18 $\left\{ {1 \atop 1\ 1} \right\} = \left\{ {2 \atop 1\ 1} \right\} = -\dfrac{\eta}{2\,\xi\,(\eta + \xi)}$;

$$\left\{ {1 \atop 1\ 2} \right\} = \left\{ {2 \atop 1\ 2} \right\} = \left\{ {1 \atop 2\ 1} \right\} = \left\{ {2 \atop 2\ 1} \right\} = \frac{1}{2(\eta + \xi)};$$

$$\left\{ {3 \atop 1\ 3} \right\} = \left\{ {3 \atop 3\ 1} \right\} = \frac{1}{2\xi};$$

$$\left\{ {1 \atop 2\ 2} \right\} = \left\{ {2 \atop 2\ 2} \right\} = -\frac{\xi}{2\,\eta(\eta + \xi)};$$

$$\left\{ {3 \atop 2\ 3} \right\} = \left\{ {3 \atop 3\ 2} \right\} = \frac{1}{2\eta};$$

$$\left\{ {1 \atop 3\ 3} \right\} = \left\{ {2 \atop 3\ 3} \right\} = -\frac{2\xi\eta}{\eta + \xi}.$$

All other $\left\{ {i \atop j\,k} \right\}$ vanish.

$[11, 1] = -\eta/8(\xi)^2$; $[11, 2] = -[12, 1] = -[21, 1] = -1/8\xi$;
$[13, 3] = [31, 3] = -[33, 1] = \eta/2$;

$[22,1] = -[12,2] = -[21,2] = -1/8\eta$; $[22,2] = -\xi/8(\eta)^2$;

$[23,3] = [32,3] = -[33,2] = \xi/2$;

All other $[ij,k]$ vanish.

7–19 $\left\{\begin{matrix} 2 \\ 1\ \ 2\end{matrix}\right\} = \left\{\begin{matrix} 2 \\ 2\ \ 1\end{matrix}\right\} = \dfrac{1}{r}$,

$\left\{\begin{matrix} 3 \\ 1\ \ 2\end{matrix}\right\} = \left\{\begin{matrix} 3 \\ 2\ \ 1\end{matrix}\right\} = -\dfrac{s}{\theta r}$,

$\left\{\begin{matrix} 1 \\ 2\ \ 2\end{matrix}\right\} = -r$,

$\left\{\begin{matrix} 3 \\ 2\ \ 3\end{matrix}\right\} = \left\{\begin{matrix} 3 \\ 3\ \ 2\end{matrix}\right\} = \dfrac{1}{\theta}$,

All other $\left\{\begin{matrix} i \\ j\ \ k\end{matrix}\right\}$ vanish.

$[12, 2] = [21,2] = r$,

$[23,2] = [32,2] = s$,

$[22,1] = -r$,

$[23,3] = [32,3] = \theta$,

All other $[ij,k]$ vanish.

7–22
$$\nabla\psi = \frac{\partial\psi}{\partial r}\,\mathbf{a}_1 + \frac{1}{(b + r\cos\chi)^2}\frac{\partial\psi}{\partial\theta}\,\mathbf{a}_2 + \frac{1}{(r)^2}\frac{\partial\psi}{\partial\chi}\,\mathbf{a}_3;$$

$$\nabla\cdot\mathbf{v} = \frac{\partial v^1}{\partial r} + \frac{\partial v^2}{\partial\theta} + \frac{\partial v^3}{\partial\chi} + \frac{v^1}{r} + \frac{v^1\cos\chi - v^3\,r\sin\chi}{b + r\cos\chi};$$

$$\nabla\Lambda\mathbf{v} = \frac{1}{r(b + r\cos\chi)}\left\{\mathbf{a}_1\left[(r)^2\frac{\partial v^3}{\partial\theta} + 2\,r\sin\chi\,(b + r\cos\chi)v^2 - (b + r\cos\chi)^2\frac{\partial v^2}{\partial\chi}\right] + \mathbf{a}_2\left[\frac{\partial v^1}{\partial\chi} - 2\,rv^3 - (r)^2\frac{\partial v^3}{\partial r}\right] + \mathbf{a}_3\left[2\cos\chi\,(b + r\cos\chi)v^2 + (b + r\cos\chi)^2\frac{\partial v^2}{\partial r} - \frac{\partial v^1}{\partial\theta}\right]\right\};$$

$$\nabla^2\psi = \frac{\partial^2\psi}{\partial r^2} + \frac{b + 2r\cos\chi}{b + r\cos\chi}\frac{\partial\psi}{\partial r} + \frac{1}{(b + r\cos\chi)^2}\frac{\partial^2\psi}{\partial\theta^2}$$

$$+ \frac{1}{(r)^2}\frac{\partial^2\psi}{\partial\chi^2} - \frac{\sin\chi}{r(b + r\cos\chi)}\frac{\partial\psi}{\partial\chi}.$$

7–23 $$\nabla\psi = \frac{4\xi}{\eta + \xi}\frac{\partial\psi}{\partial\xi}\mathbf{a}_1 + \frac{4\eta}{\eta + \xi}\frac{\partial\psi}{\partial\eta}\mathbf{a}_2 + \frac{1}{\xi\eta}\frac{\partial\psi}{\partial\phi}\mathbf{a}_3$$

$$\nabla \cdot \mathbf{v} = \frac{\partial v^1}{\partial\xi} + \frac{\partial v^2}{\partial\eta} + \frac{\partial v^3}{\partial\phi} + \frac{v^1 + v^2}{\eta + \xi};$$

$$\nabla \Lambda \mathbf{v} = \frac{4}{\eta + \xi}\left\{\mathbf{a}_1\left[\xi v^3 + \xi\eta\frac{\partial v^3}{\partial\eta} - \frac{\eta + \xi}{4\eta}\frac{\partial v^2}{\partial\phi}\right]\right.$$

$$+ \mathbf{a}_2\left[\frac{\eta + \xi}{4\xi}\frac{\partial v^1}{\partial\phi} - \eta\xi\frac{\partial v^3}{\partial\xi}\right]$$

$$\left. + \mathbf{a}_3\left[\frac{\eta + \xi}{4\eta}\frac{\partial v^2}{\partial\xi} - \frac{\eta + \xi}{4\xi}\frac{\partial v^1}{\partial\eta} - \frac{v^1}{4\xi} + \frac{v^2}{4\eta}\right]\right\};$$

$$\nabla^2\psi = \frac{4}{\eta + \xi}\left[\frac{\partial\psi}{\partial\xi} + \xi\frac{\partial^2\psi}{\partial\xi^2} + \frac{\partial\psi}{\partial\eta} + \eta\frac{\partial^2\psi}{\partial\eta^2} + \frac{\eta + \xi}{4\xi\eta}\frac{\partial^2\psi}{\partial\phi^2}\right].$$

7–24 $$\nabla\psi = \mathbf{a}_1\frac{\partial\psi}{\partial r} + \frac{\mathbf{a}_2}{(r)^2}\left[\frac{\partial\psi}{\partial\theta} - \frac{s}{\theta}\frac{\partial\psi}{\partial s}\right] + \frac{\mathbf{a}_3}{(r)^2\theta}\left[\frac{r^2 + s^2}{\theta}\frac{\partial\psi}{\partial s} - s\frac{\partial\psi}{\partial\theta}\right];$$

$$\nabla \cdot \mathbf{v} = \frac{\partial v^1}{\partial r} + \frac{\partial v^2}{\partial\theta} + \frac{\partial v^3}{\partial s} + \frac{v^1}{r} + \frac{v^2}{\theta};$$

$$\nabla \Lambda \mathbf{v} = \frac{1}{\theta r}\left\{\left[\theta v^3 + s\theta\frac{\partial v^2}{\partial\theta} + \theta^2\frac{\partial v^3}{\partial\theta} - sv^2 - (r^2 + s^2)\frac{\partial v^2}{\partial s}\right.\right.$$

$$\left. - s\theta\frac{\partial v^3}{\partial s}\right]\mathbf{a}_1$$

$$+ \left[\frac{\partial v^1}{\partial s} - s\theta\frac{\partial v^2}{\partial r} - \theta^2\frac{\partial v^3}{\partial r}\right]\mathbf{a}_2$$

$$\left. + \left[2rv^2 + (r^2 + s^2)\frac{\partial v^2}{\partial r} + s\theta\frac{\partial v^3}{\partial r} - \frac{\partial v^1}{\partial\theta}\right]\mathbf{a}_3\right\}$$

$$\nabla^2\psi = \frac{\partial^2\psi}{\partial r^2} + \frac{1}{r}\frac{\partial\psi}{\partial r} + \frac{r^2 + s^2}{r^2\,\theta^2}\frac{\partial^2\psi}{\partial s^2} + \frac{2s}{r^2\,\theta^2}\frac{\partial\psi}{\partial s}$$

$$+ \frac{1}{r^2}\frac{\partial^2\psi}{\partial\theta^2} - \frac{2s}{\theta r^2}\frac{\partial^2\psi}{\partial\,\theta\partial s}\,.$$

7-25 $$\nabla\psi = g^{ij}\frac{\partial\psi}{\partial x^j}\,\mathbf{a}_i\,;$$

$$\nabla\cdot\mathbf{v} = \frac{\partial v^i}{\partial x^i}\,;$$

$$\nabla\wedge\mathbf{v} = \frac{1}{V}\left\{\left[g_{3i}\frac{\partial v^i}{\partial x^2} - g_{2i}\frac{\partial v^i}{\partial x^3}\right]\mathbf{a}_1 + \left[g_{1i}\frac{\partial v^i}{\partial x^3} - g_{3i}\frac{\partial v^i}{\partial x^1}\right]\mathbf{a}_2 + \left[g_{2i}\frac{\partial v^i}{\partial x^1} - g_{1i}\frac{\partial v^i}{\partial x^2}\right]\mathbf{a}_3\right\};$$

$$\nabla^2\psi = g^{ij}\frac{\partial^2\psi}{\partial x^i\,\partial x^j}\,.$$

7-26 Momentum components are covariant; $p_i\,q^i$ and $dp_i\,dq^i$ are invariant.

Index

A CATALOG OF SELECTED

DOVER BOOKS

IN SCIENCE AND MATHEMATICS

A CATALOG OF SELECTED

DOVER BOOKS

IN SCIENCE AND MATHEMATICS

QUALITATIVE THEORY OF DIFFERENTIAL EQUATIONS, V.V. Nemytskii and V.V. Stepanov. Classic graduate-level text by two prominent Soviet mathematicians covers classical differential equations as well as topological dynamics and ergodic theory. Bibliographies. 523pp. 5⅜ × 8½. 65954-2 Pa. $10.95

MATRICES AND LINEAR ALGEBRA, Hans Schneider and George Phillip Barker. Basic textbook covers theory of matrices and its applications to systems of linear equations and related topics such as determinants, eigenvalues and differential equations. Numerous exercises. 432pp. 5⅜ × 8½. 66014-1 Pa. $10.95

QUANTUM THEORY, David Bohm. This advanced undergraduate-level text presents the quantum theory in terms of qualitative and imaginative concepts, followed by specific applications worked out in mathematical detail. Preface. Index. 655pp. 5⅜ × 8½. 65969-0 Pa. $13.95

ATOMIC PHYSICS (8th edition), Max Born. Nobel laureate's lucid treatment of kinetic theory of gases, elementary particles, nuclear atom, wave-corpuscles, atomic structure and spectral lines, much more. Over 40 appendices, bibliography. 495pp. 5⅜ × 8½. 65984-4 Pa. $12.95

ELECTRONIC STRUCTURE AND THE PROPERTIES OF SOLIDS: The Physics of the Chemical Bond, Walter A. Harrison. Innovative text offers basic understanding of the electronic structure of covalent and ionic solids, simple metals, transition metals and their compounds. Problems. 1980 edition. 582pp. 6⅛ × 9¼. 66021-4 Pa. $15.95

BOUNDARY VALUE PROBLEMS OF HEAT CONDUCTION, M. Necati Özisik. Systematic, comprehensive treatment of modern mathematical methods of solving problems in heat conduction and diffusion. Numerous examples and problems. Selected references. Appendices. 505pp. 5⅜ × 8½. 65990-9 Pa. $12.95

A SHORT HISTORY OF CHEMISTRY (3rd edition), J.R. Partington. Classic exposition explores origins of chemistry, alchemy, early medical chemistry, nature of atmosphere, theory of valency, laws and structure of atomic theory, much more. 428pp. 5⅜ × 8½. (Available in U.S. only) 65977-1 Pa. $10.95

A HISTORY OF ASTRONOMY, A. Pannekoek. Well-balanced, carefully reasoned study covers such topics as Ptolemaic theory, work of Copernicus, Kepler, Newton, Eddington's work on stars, much more. Illustrated. References. 521pp. 5⅜ × 8½. 65994-1 Pa. $12.95

PRINCIPLES OF METEOROLOGICAL ANALYSIS, Walter J. Saucier. Highly respected, abundantly illustrated classic reviews atmospheric variables, hydrostatics, static stability, various analyses (scalar, cross-section, isobaric, isentropic, more). For intermediate meteorology students. 454pp. 6⅛ × 9¼. 65979-8 Pa. $14.95

NUMERICAL METHODS FOR SCIENTISTS AND ENGINEERS, Richard Hamming. Classic text stresses frequency approach in coverage of algorithms, polynomial approximation, Fourier approximation, exponential approximation, other topics. Revised and enlarged 2nd edition. 721pp. 5⅜ × 8½.
65241-6 Pa. $14.95

THEORETICAL SOLID STATE PHYSICS, Vol. I: Perfect Lattices in Equilibrium; Vol. II: Non-Equilibrium and Disorder, William Jones and Norman H. March. Monumental reference work covers fundamental theory of equilibrium properties of perfect crystalline solids, non-equilibrium properties, defects and disordered systems. Appendices. Problems. Preface. Diagrams. Index. Bibliography. Total of 1,301pp. 5⅜ × 8½. Two volumes. Vol. I 65015-4 Pa. $14.95
Vol. II 65016-2 Pa. $14.95

OPTIMIZATION THEORY WITH APPLICATIONS, Donald A. Pierre. Broad-spectrum approach to important topic. Classical theory of minima and maxima, calculus of variations, simplex technique and linear programming, more. Many problems, examples. 640pp. 5⅜ × 8½. 65205-X Pa. $14.95

THE CONTINUUM: A Critical Examination of the Foundation of Analysis, Hermann Weyl. Classic of 20th-century foundational research deals with the conceptual problem posed by the continuum. 156pp. 5⅜ × 8½. 67982-9 Pa. $5.95

ESSAYS ON THE THEORY OF NUMBERS, Richard Dedekind. Two classic essays by great German mathematician: on the theory of irrational numbers; and on transfinite numbers and properties of natural numbers. 115pp. 5⅜ × 8½.
21010-3 Pa. $4.95

THE FUNCTIONS OF MATHEMATICAL PHYSICS, Harry Hochstadt. Comprehensive treatment of orthogonal polynomials, hypergeometric functions, Hill's equation, much more. Bibliography. Index. 322pp. 5⅜ × 8½. 65214-9 Pa. $9.95

NUMBER THEORY AND ITS HISTORY, Oystein Ore. Unusually clear, accessible introduction covers counting, properties of numbers, prime numbers, much more. Bibliography. 380pp. 5⅜ × 8½. 65620-9 Pa. $9.95

THE VARIATIONAL PRINCIPLES OF MECHANICS, Cornelius Lanczos. Graduate level coverage of calculus of variations, equations of motion, relativistic mechanics, more. First inexpensive paperbound edition of classic treatise. Index. Bibliography. 418pp. 5⅜ × 8½. 65067-7 Pa. $11.95

MATHEMATICAL TABLES AND FORMULAS, Robert D. Carmichael and Edwin R. Smith. Logarithms, sines, tangents, trig functions, powers, roots, reciprocals, exponential and hyperbolic functions, formulas and theorems. 269pp. 5⅜ × 8½. 60111-0 Pa. $6.95

THEORETICAL PHYSICS, Georg Joos, with Ira M. Freeman. Classic overview covers essential math, mechanics, electromagnetic theory, thermodynamics, quantum mechanics, nuclear physics, other topics. First paperback edition. xxiii + 885pp. 5⅜ × 8½. 65227-0 Pa. $19.95

HANDBOOK OF MATHEMATICAL FUNCTIONS WITH FORMULAS, GRAPHS, AND MATHEMATICAL TABLES, edited by Milton Abramowitz and Irene A. Stegun. Vast compendium: 29 sets of tables, some to as high as 20 places. 1,046pp. 8 × 10½. 61272-4 Pa. $24.95

MATHEMATICAL METHODS IN PHYSICS AND ENGINEERING, John W. Dettman. Algebraically based approach to vectors, mapping, diffraction, other topics in applied math. Also generalized functions, analytic function theory, more. Exercises. 448pp. 5⅜ × 8¼. 65649-7 Pa. $9.95

A SURVEY OF NUMERICAL MATHEMATICS, David M. Young and Robert Todd Gregory. Broad self-contained coverage of computer-oriented numerical algorithms for solving various types of mathematical problems in linear algebra, ordinary and partial, differential equations, much more. Exercises. Total of 1,248pp. 5⅜ × 8½. Two volumes. Vol. I 65691-8 Pa. $14.95
Vol. II 65692-6 Pa. $14.95

TENSOR ANALYSIS FOR PHYSICISTS, J.A. Schouten. Concise exposition of the mathematical basis of tensor analysis, integrated with well-chosen physical examples of the theory. Exercises. Index. Bibliography. 289pp. 5⅜ × 8½. 65582-2 Pa. $8.95

INTRODUCTION TO NUMERICAL ANALYSIS (2nd Edition), F.B. Hildebrand. Classic, fundamental treatment covers computation, approximation, interpolation, numerical differentiation and integration, other topics. 150 new problems. 669pp. 5⅜ × 8½. 65363-3 Pa. $15.95

INVESTIGATIONS ON THE THEORY OF THE BROWNIAN MOVEMENT, Albert Einstein. Five papers (1905–8) investigating dynamics of Brownian motion and evolving elementary theory. Notes by R. Fürth. 122pp. 5⅜ × 8½. 60304-0 Pa. $4.95

CATASTROPHE THEORY FOR SCIENTISTS AND ENGINEERS, Robert Gilmore. Advanced-level treatment describes mathematics of theory grounded in the work of Poincaré, R. Thom, other mathematicians. Also important applications to problems in mathematics, physics, chemistry and engineering. 1981 edition. References. 28 tables. 397 black-and-white illustrations. xvii + 666pp. 6⅛ × 9¼. 67539-4 Pa. $16.95

AN INTRODUCTION TO STATISTICAL THERMODYNAMICS, Terrell L. Hill. Excellent basic text offers wide-ranging coverage of quantum statistical mechanics, systems of interacting molecules, quantum statistics, more. 523pp. 5⅜ × 8½. 65242-4 Pa. $12.95

ELEMENTARY DIFFERENTIAL EQUATIONS, William Ted Martin and Eric Reissner. Exceptionally clear, comprehensive introduction at undergraduate level. Nature and origin of differential equations, differential equations of first, second and higher orders. Picard's Theorem, much more. Problems with solutions. 331pp. 5⅜ × 8½. 65024-3 Pa. $8.95

STATISTICAL PHYSICS, Gregory H. Wannier. Classic text combines thermodynamics, statistical mechanics and kinetic theory in one unified presentation of thermal physics. Problems with solutions. Bibliography. 532pp. 5⅜ × 8½. 65401-X Pa. $12.95

ROTARY-WING AERODYNAMICS, W.Z. Stepniewski. Clear, concise text covers aerodynamic phenomena of the rotor and offers guidelines for helicopter performance evaluation. Originally prepared for NASA. 537 figures. 640pp. 6⅛ × 9¼.
64647-5 Pa. $15.95

DIFFERENTIAL GEOMETRY, Heinrich W. Guggenheimer. Local differential geometry as an application of advanced calculus and linear algebra. Curvature, transformation groups, surfaces, more. Exercises. 62 figures. 378pp. 5⅜ × 8½.
63433-7 Pa. $8.95

INTRODUCTION TO SPACE DYNAMICS, William Tyrrell Thomson. Comprehensive, classic introduction to space-flight engineering for advanced undergraduate and graduate students. Includes vector algebra, kinematics, transformation of coordinates. Bibliography. Index. 352pp. 5⅜ × 8½. 65113-4 Pa. $8.95

A SURVEY OF MINIMAL SURFACES, Robert Osserman. Up-to-date, in-depth discussion of the field for advanced students. Corrected and enlarged edition covers new developments. Includes numerous problems. 192pp. 5⅜ × 8½.
64998-9 Pa. $8.95

ANALYTICAL MECHANICS OF GEARS, Earle Buckingham. Indispensable reference for modern gear manufacture covers conjugate gear-tooth action, gear-tooth profiles of various gears, many other topics. 263 figures. 102 tables. 546pp. 5⅜ × 8½. 65712-4 Pa. $14.95

SET THEORY AND LOGIC, Robert R. Stoll. Lucid introduction to unified theory of mathematical concepts. Set theory and logic seen as tools for conceptual understanding of real number system. 496pp. 5⅜ × 8¼. 63829-4 Pa. $12.95

A HISTORY OF MECHANICS, René Dugas. Monumental study of mechanical principles from antiquity to quantum mechanics. Contributions of ancient Greeks, Galileo, Leonardo, Kepler, Lagrange, many others. 671pp. 5⅜ × 8½.
65632-2 Pa. $14.95

FAMOUS PROBLEMS OF GEOMETRY AND HOW TO SOLVE THEM, Benjamin Bold. Squaring the circle, trisecting the angle, duplicating the cube: learn their history, why they are impossible to solve, then solve them yourself. 128pp. 5⅜ × 8½. 24297-8 Pa. $4.95

MECHANICAL VIBRATIONS, J.P. Den Hartog. Classic textbook offers lucid explanations and illustrative models, applying theories of vibrations to a variety of practical industrial engineering problems. Numerous figures. 233 problems, solutions. Appendix. Index. Preface. 436pp. 5⅜ × 8½. 64785-4 Pa. $10.95

CURVATURE AND HOMOLOGY, Samuel I. Goldberg. Thorough treatment of specialized branch of differential geometry. Covers Riemannian manifolds, topology of differentiable manifolds, compact Lie groups, other topics. Exercises. 315pp. 5⅜ × 8½. 64314-X Pa. $9.95

HISTORY OF STRENGTH OF MATERIALS, Stephen P. Timoshenko. Excellent historical survey of the strength of materials with many references to the theories of elasticity and structure. 245 figures. 452pp. 5⅜ × 8½. 61187-6 Pa. $11.95

THE FOUR-COLOR PROBLEM: Assaults and Conquest, Thomas L. Saaty and Paul G. Kainen. Engrossing, comprehensive account of the century-old combinatorial topological problem, its history and solution. Bibliographies. Index. 110 figures. 228pp. 5⅜ × 8½. 65092-8 Pa. $6.95

CATALYSIS IN CHEMISTRY AND ENZYMOLOGY, William P. Jencks. Exceptionally clear coverage of mechanisms for catalysis, forces in aqueous solution, carbonyl- and acyl-group reactions, practical kinetics, more. 864pp. 5⅜ × 8½. 65460-5 Pa. $19.95

PROBABILITY: An Introduction, Samuel Goldberg. Excellent basic text covers set theory, probability theory for finite sample spaces, binomial theorem, much more. 360 problems. Bibliographies. 322pp. 5⅜ × 8½. 65252-1 Pa. $8.95

LIGHTNING, Martin A. Uman. Revised, updated edition of classic work on the physics of lightning. Phenomena, terminology, measurement, photography, spectroscopy, thunder, more. Reviews recent research. Bibliography. Indices. 320pp. 5⅜ × 8¼. 64575-4 Pa. $8.95

PROBABILITY THEORY: A Concise Course, Y.A. Rozanov. Highly readable, self-contained introduction covers combination of events, dependent events, Bernoulli trials, etc. Translation by Richard Silverman. 148pp. 5⅜ × 8¼. 63544-9 Pa. $5.95

AN INTRODUCTION TO HAMILTONIAN OPTICS, H. A. Buchdahl. Detailed account of the Hamiltonian treatment of aberration theory in geometrical optics. Many classes of optical systems defined in terms of the symmetries they possess. Problems with detailed solutions. 1970 edition. xv + 360pp. 5⅜ × 8½. 67597-1 Pa. $10.95

STATISTICS MANUAL, Edwin L. Crow, et al. Comprehensive, practical collection of classical and modern methods prepared by U.S. Naval Ordnance Test Station. Stress on use. Basics of statistics assumed. 288pp. 5⅜ × 8½. 60599-X Pa. $6.95

DICTIONARY/OUTLINE OF BASIC STATISTICS, John E. Freund and Frank J. Williams. A clear concise dictionary of over 1,000 statistical terms and an outline of statistical formulas covering probability, nonparametric tests, much more. 208pp. 5⅜ × 8½. 66796-0 Pa. $6.95

STATISTICAL METHOD FROM THE VIEWPOINT OF QUALITY CONTROL, Walter A. Shewhart. Important text explains regulation of variables, uses of statistical control to achieve quality control in industry, agriculture, other areas. 192pp. 5⅜ × 8½. 65232-7 Pa. $7.95

THE INTERPRETATION OF GEOLOGICAL PHASE DIAGRAMS, Ernest G. Ehlers. Clear, concise text emphasizes diagrams of systems under fluid or containing pressure; also coverage of complex binary systems, hydrothermal melting, more. 288pp. 6½ × 9¼. 65389-7 Pa. $10.95

STATISTICAL ADJUSTMENT OF DATA, W. Edwards Deming. Introduction to basic concepts of statistics, curve fitting, least squares solution, conditions without parameter, conditions containing parameters. 26 exercises worked out. 271pp. 5⅜ × 8½. 64685-8 Pa. $8.95

CATALOG OF DOVER BOOKS

CHALLENGING MATHEMATICAL PROBLEMS WITH ELEMENTARY SOLUTIONS, A.M. Yaglom and I.M. Yaglom. Over 170 challenging problems on probability theory, combinatorial analysis, points and lines, topology, convex polygons, many other topics. Solutions. Total of 445pp. 5⅜ × 8½. Two-vol. set.
Vol. I 65536-9 Pa. $7.95
Vol. II 65537-7 Pa. $6.95

FIFTY CHALLENGING PROBLEMS IN PROBABILITY WITH SOLUTIONS, Frederick Mosteller. Remarkable puzzlers, graded in difficulty, illustrate elementary and advanced aspects of probability. Detailed solutions. 88pp. 5⅜ × 8½.
65355-2 Pa. $4.95

EXPERIMENTS IN TOPOLOGY, Stephen Barr. Classic, lively explanation of one of the byways of mathematics. Klein bottles, Moebius strips, projective planes, map coloring, problem of the Koenigsberg bridges, much more, described with clarity and wit. 43 figures. 210pp. 5⅜ × 8½. 25933-1 Pa. $5.95

RELATIVITY IN ILLUSTRATIONS, Jacob T. Schwartz. Clear nontechnical treatment makes relativity more accessible than ever before. Over 60 drawings illustrate concepts more clearly than text alone. Only high school geometry needed. Bibliography. 128pp. 6⅛ × 9¼. 25965-X Pa. $6.95

AN INTRODUCTION TO ORDINARY DIFFERENTIAL EQUATIONS, Earl A. Coddington. A thorough and systematic first course in elementary differential equations for undergraduates in mathematics and science, with many exercises and problems (with answers). Index. 304pp. 5⅜ × 8½. 65942-9 Pa. $8.95

FOURIER SERIES AND ORTHOGONAL FUNCTIONS, Harry F. Davis. An incisive text combining theory and practical example to introduce Fourier series, orthogonal functions and applications of the Fourier method to boundary-value problems. 570 exercises. Answers and notes. 416pp. 5⅜ × 8½. 65973-9 Pa. $9.95

THE THEORY OF BRANCHING PROCESSES, Theodore E. Harris. First systematic, comprehensive treatment of branching (i.e. multiplicative) processes and their applications. Galton-Watson model, Markov branching processes, electron-photon cascade, many other topics. Rigorous proofs. Bibliography. 240pp. 5⅜ × 8½. 65952-6 Pa. $6.95

AN INTRODUCTION TO ALGEBRAIC STRUCTURES, Joseph Landin. Superb self-contained text covers "abstract algebra": sets and numbers, theory of groups, theory of rings, much more. Numerous well-chosen examples, exercises. 247pp. 5⅜ × 8½. 65940-2 Pa. $7.95

Prices subject to change without notice.

Available at your book dealer or write for free Mathematics and Science Catalog to Dept. GI, Dover Publications, Inc., 31 East 2nd St., Mineola, N.Y. 11501. Dover publishes more than 175 books each year on science, elementary and advanced mathematics, biology, music, art, literature, history, social sciences and other areas.